SYNTHESES AND PHYSICAL STUDIES

OF

INORGANIC COMPOUNDS

SYNTHESES AND PHYSICAL STUDIES OF INORGANIC COMPOUNDS

BY

C. F. BELL, M.A., D.PHIL., F.R.I.C.

Lecturer in Inorganic Chemistry
Brunel University, London

PERGAMON PRESS

OXFORD · NEW YORK
TORONTO · SYDNEY · BRAUNSCHWEIG

CHEMISTRY

Pergamon Press Ltd., Headington Hill Hall, Oxford
Pergamon Press Inc., Maxwell House, Fairview Park, Elmsford,
New York 10523
Pergamon of Canada Ltd., 207 Queen's Quay West, Toronto 1
Pergamon Press (Aust.) Pty. Ltd., 19a Boundary Street,
Rushcutters Bay, N.S.W. 2011, Australia
Vieweg & Sohn GmbH, Burgplatz 1, Braunschweig

First edition 1972

Library of Congress Catalog Card No. 79-178772

Printed in Germany
08   016651 2

TO

SHEILA

9995

CONTENTS

PREFACE

THE renaissance of inorganic chemistry during the past three decades owes much of its origin and continuing impetus to advances in three directions. Firstly, existing preparative techniques have been improved and new ones developed, resulting in the discovery of many new kinds of inorganic compounds. Secondly, established methods for studying the physico-chemical properties and structures of chemical compounds have been supplemented by a number of important, recently developed techniques, notably nuclear magnetic resonance, electron spin resonance and Mössbauer absorption spectroscopy. Thirdly, progress in theoretical chemistry, particularly in the evolution of the molecular orbital and ligand field theories, has facilitated the rationalization of much experimental data and deepened our understanding of the forces which bind atoms together.

The present state of inorganic chemistry is reflected in the research publications appearing in the literature. Many current papers are devoted to establishing the detailed physical and chemical properties of specific compounds with the emphasis very much on the application of physical principles and investigational techniques and the theoretical interpretation of experimental data.

Concurrently with the great expansion of research activity in inorganic chemistry, the teaching of this subject has veered sharply towards comparative treatments of the chemistry of the elements based on the principles underlying the Periodic Table. In such treatments, the student gains an overall appreciation of inorganic chemistry without being burdened with excessive detail. There still remains, however, a significant gap between this approach and many of the topics which excite the interest of research chemists today.

There are various ways in which recent advances can be presented to the undergraduate or postgraduate student. One is to adopt the approach exemplified in two recent books (Drago, R., *Physical Methods in Inorganic Chemistry*, Reinhold, New York, 1965, and Hill, H. A. O., and Day, P., *Physical Methods in Advanced Inorganic Chemistry*, Interscience, London, 1968) on physical methods as applied to inorganic chemistry. The emphasis in this approach is on the principles and experimentation of the methods discussed and undoubtedly it provides a sound foundation for the advanced study of the subject.

In this book I have chosen an alternative approach. This is based on the consideration, in some depth, of the synthesis, properties and reactions and structures of a number of compounds, selected on the criterion that the study of each has resulted in important contributions to our present practice and understanding of inorganic chemistry. Details of experimental procedures are, in general, not included, for these may be found elsewhere. For example, the syntheses of almost all the compounds discussed in the following pages are described in full in the appropriate volume of the series *Inorganic Syntheses* (McGraw-Hill).

My aim throughout has been to combine the results obtained in many different physico-chemical investigations of each compound in order to show the current state of our knowledge. This approach also serves to illustrate how closely interrelated and interdependent inorganic and physical chemistry have now become and the artificiality of trying to maintain sharplydefined boundaries between the two disciplines. I trust that this book will introduce the reader to many of the significant research papers of recent years and will encourage him to pursue his own studies of the chemical literature.

In the text, the S.I. units are used for the expression of most physical data. There are the following exceptions. In the cases of magnetic and dipole moments, traditional units are preferred because of the awkward magnitudes of the S.I. units. The S.I. equivalent of each is given in the text on the occasion of the first reference to it and thereafter the traditional unit is used. Temperatures are quoted in degrees Celsius, °C, and wave-numbers in reciprocal centimetres, cm^{-1}. Althoug neither is strictly in accordance with S.I. practice, these are still the units in common usage and it is convenient to retain them here for ease of crossreference with the literature.

I wish to record my grateful thanks to the authors and publishers who have given permission to reproduce figures as indicated in the text and to all those research chemists without whose original inspiration and creative work this book would never have been possible.

I would like especially to thank Professor H. M. N. H. Irving for his original idea regarding the writing of this book and for his continued interest and help throughout its preparation.

Finally, I must record my gratitude to my wife for her considerable help in the typing and editing of this work and my deep appreciation of her patience and encouragement in bringing it to completion.

CHAPTER 1

DIBORANE

Introduction

The first observation of compound formation between boron and hydrogen appears to have been made by Sir Humphrey Davy[1] in 1809. He noted that a gas was obtained when the product of the reaction of boron trioxide with potassium was treated with water or dilute hydrochloric acid. This was mainly hydrogen, but also contained another component which burnt with a blue flame tinged with green. He identified this as a volatile compound of boron and hydrogen.

From time to time during the next 100 years, hydrides of boron were made by other research workers. These compounds were not correctly identified, probably because of the difficulty in purifying and analysing such reactive substances.

Alfred Stock and his co-workers were the first to characterize the hydrides of boron (or boranes as they are now commonly called). Their work[2] is remarkable for the development of experimental techniques for the handling of highly reactive, volatile compounds. They were able to establish the existence of as many as ten boranes. Stock's work provides the basis for modern synthetic practice and it remains one of the great, classical contributions to the development of inorganic chemistry in this century.

For many years, interest in borane chemistry was stimulated by the recognition that, in their stoichiometry, these compounds did not conform with accepted theories of valence. They, and some other similar inorganic molecules, are 'electron-deficient' because there are insufficient valence electrons available to make conventional two-electron bonds between each pair of atoms bonded together. It has become necessary to introduce the concept of a multicentre bond into valence theory to account for the bonding in such compounds.

More recently, boranes have been intensively studied in view of their potentialities as chemical fuels. From a consideration of energy content per unit weight, boranes are theoretically much superior to hydrocarbons as fuels, although the difficulty of handling them in large quantities has impeded their exploitation on an industrial scale. Some borane derivatives, notably the tetrahydroborates, are applied to an increasing extent as reducing agents for both inorganic and organic compounds.

As the simplest known borane, diborane (B_2H_6) has attracted special interest and it has been the subject of very many physico-chemical studies.

Preparation

Stock prepared a mixture of tetraborane, B_4H_{10}, pentaborane, B_5H_9, hexaborane, B_6H_{10}, and decaborane, $B_{10}H_{14}$, by the reaction between magnesium boride and hydrochloric acid. Separation from impurities such as carbon dioxide, hydrogen sulphide and silanes proved difficult and yields were low. Better yields were obtained by the use of phosphoric instead of hydrochloric acid. Diborane itself is rapidly decomposed in aqueous solution and so is not formed directly in the above reaction and it was made by Stock by the pyrolysis of B_4H_{10}.

Schlesinger and Burg[3] greatly improved the synthesis of B_2H_6 by the action of an electric discharge on a mixture of BCl_3 and excess hydrogen at low pressure. Monochlorodiborane, B_2H_5Cl, is formed. This disproportionates at $0°C$ and atmospheric pressure into B_2H_6 and BCl_3.

Neither of these reactions is convenient in practice and they have been superseded[4] by the reduction of BCl_3 or BF_3 with lithium aluminium hydride, $LiAlH_4$.

$$3LiAlH_4 + 4BCl_3 \rightarrow 3LiCl + 3AlCl_3 + 2B_2H_6$$

$LiAlH_4$ is dissolved in diethyl ether and allowed to react with the diethyl ether complex of BCl_3, $(C_2H_5)_2O \cdot BCl_3$, at a low temperature. B_2H_6 is released when the reaction mixture is warmed to room temperature.

A method recommended[5,6] for the preparation of small quantities of diborane involves the reaction of potassium tetrahydroborate with 85% ortho-phosphoric acid.

$$2KBH_4 + 3H_3PO_4 \rightarrow B_2H_6 + 2H_2 + 2KH_2PO_4$$

The reaction is carried out *in vacuo* at room temperature and the volatile products led into a cold trap at $-196°C$. High-purity diborane, m.p. $-165°$, b.p. $-90°C$, in 40–50% yield, is produced.

Better yields of B_2H_6 result from the reaction of a metal tetrahydroborate with halides of bismuth, tin, mercury or antimony.[7] For example, up to 98% yield has been achieved by the reaction between sodium tetrahydroborate and mercury(I) chloride, both in solution in 'diglyme' (the dimethylether of diethyleneglycol), at room temperature.

$$2NaBH_4 + Hg_2Cl_2 \rightarrow 2Hg + 2NaCl + B_2H_6 + H_2$$

Halides of other metals are less efficient. When $SbCl_3$ is used, some stibine contaminates the product and with $PbCl_2$ only 20% yield is achieved.

The synthesis of diborane directly from readily available materials would have clear advantages over earlier syntheses, such as those requiring the use of highly reactive boron halides. B_2H_6 has been made[8] from boron trioxide, B_2O_3, by reaction with hydrogen under high pressure at $175°C$ in the presence of aluminium and aluminium chloride. Conversions of up to 50% have been achieved. Another synthetic route is the low-pressure hydrogenation of boron monoxide, B_2O_2, at $1200°C$.[9] B_2O_2 is made in the pure state by the reaction between TiO_2 or ZrO_2 and boron carbide, B_4C, *in vacuo*. The major product of this hydrogenation reaction is B_2H_6 in yields of up to 20%.

Reactions

Diborane, like the other boranes, is a highly reactive compound. It is thermally unstable and decomposes rapidly at $100°C$ to give hydrogen and higher boranes. It is readily oxidized by air but does not, unlike some of the other boranes, inflame spontaneously in air.

When B_2H_6 comes into contact with water, immediate hydrolysis occurs.

$$B_2H_6 + 6H_2O \rightarrow 2H_3BO_3 + 6H_2$$

One or two hydrogen atoms can be directly replaced by halogen.

$$B_2H_6 + Br_2 \rightarrow B_2H_5Br + HBr$$

B_2H_5Cl can be made in a similar way, but it is difficult to isolate because of some dispro-portionation into BCl_3 and B_2H_6 at room temperature. The reaction between B_2H_6 and BCl_3 gives,[10] at room temperature, a mixture of $BHCl_2$ and B_2H_5Cl, but at 100°C the product is almost exclusively $BHCl_2$. This indicates that the replacement of a second hy-drogen atom by halogen causes a cleavage of the B_2H_6 molecule into two parts.

There is no evidence for the independent existence of the borane radical, BH_3, although in many of its reactions diborane behaves as two BH_3 fragments. For example, B_2H_6 reacts with CO under high pressure at 100°C to give borane carbonyl, $BH_3 \cdot CO$. This adduct dissociates into CO and B_2H_6 at ordinary temperature and pressure. Reaction of B_2H_6 with excess trimethylamine gives trimethylamine borane, $(CH_3)_3N \cdot BH_3$. In this and the carbonyl, the BH_3 fragment acts as a Lewis acid in combination with an electron donor (Lewis base) molecule. In both reactions, the B_2H_6 molecule undergoes a symmetrical cleavage. In the case of the reaction with trimethylamine, an intermediate of composition $(CH_3)_3N \cdot B_2H_6$ has been isolated,[11] indicating that the diborane molecule is coordinated by one donor molecule before cleavage occurs.

μ-Dimethylaminodiborane, $(CH_3)_2NB_2H_5$, has been made[12] by the interaction of sodium dimethylamidotrihydroborate, $Na(CH_3)_2NBH_3$, with diborane in diglyme solution.

$$(CH_3)_2NBH_3^- + B_2H_6 \rightarrow \mu\text{-}(CH_3)_2NB_2H_5 + BH_4^-$$

The structure of this derivative has been established by electron diffraction. Like B_2H_6, it has a bridge structure (see below) in which a dimethylamino group has replaced one of the bridge hydrogen atoms of diborane.

Different products arise from the reaction between B_2H_6 and ammonia, depending on the experimental conditions. At $-120°C$, reaction with excess ammonia forms a salt-like solid of stoichiometry, $B_2H_6 \cdot 2NH_3$. This contains BH_4^- ions and is formulated as an ionic compound, $[H_2B(NH_3)_2]^+[BH_4]^-$. It results from the unsymmetrical cleavage of the B_2H_6 molecule.

When diborane is heated with excess ammonia, boron imide and, finally, boron nitride are produced. If B_2H_6 and NH_3 in a mole ration of 1:2 are heated together, or if $[H_2B(NH_3)_2]^+[BH_4]^-$ is heated above 200°C, the cyclic compound borazine, $B_3N_3H_6$, is formed.

The B_2H_6 molecule is also split unsymmetrically by dimethylsulphoxide.[13] Their reaction in a mole ratio of $B_2H_6:(CH_3)_2SO = 1:2$ in dichloromethane solution at $-78°C$ gives $BH_2[OS(CH_3)_2]_2^+[BH_4]^-$.

Reactions between diborane and metal alkyls produces metal tetrahydroborates. Thus B_2H_6 and ethyllithium react at room temperature to form lithium tetrahydroborate, $LiBH_4$.

$$2LiC_2H_5 + 2B_2H_6 \rightarrow 2LiBH_4 + (C_2H_5)_2B_2H_4$$

The BH_4^- ion can be regarded as formed by combination between the Lewis acid, BH_3, and the base, H^-. Sodium and lithium tetrahydroborates are used as selective reducing agents for organic compounds. They will reduce carbonyl groups in aldehydes and ketones to alcohols, but not those in acids, esters or acid anhydrides.

3

The great reactivity of B_2H_6 and, in many cases, the products of its reactions are consistent with a cleavage of the molecule by a process which requires comparatively little energy. The chemistry of diborane can be most readily understood from a knowledge of its structure, which has now been firmly established in a series of important investigations using a variety of physical techniques.

Structure

Structures which have been proposed from time to time for molecular diborane are essentially of three kinds:[14] (a) an analogue of ethane, $H_3B \cdot BH_3$; (b) a bridged molecule, $H_2B(H_2)BH_2$, in which two of the hydrogens act as bridge atoms between the borons, the remaining four being equivalent to one another but not to the bridge hydrogens; (c) an ionic structure, $[H_2\bar{B}=\bar{B}H_2][H^+]_2$, in which the two different kinds of hydrogen are bound respectively by covalent and ionic bonds.

Early X-ray[15] and electron diffraction[16] results were interpreted in terms of an ethanelike structure for diborane. Interatomic distances found by electron diffraction were:

B, B = 0·186 nm (cf. C, C in ethane = 0·154 nm) and B, H = 0·127 nm.

The experimental data are equally well in agreement with a bridged structure[17,18] so no conclusive structural assignment can be made from these results.

The vibrational spectrum of diborane (see below) clearly favoured a bridged structure, so it was necessary for the electron diffraction investigation to be repeated.[19] In this study, the most satisfactory agreement between theoretical radial distribution curves and those calculated from a visual examination of the electron diffraction photographs was found for the bridged model (b) (Fig. 1.1). Table I summarizes the data of Hedberg and Scho-

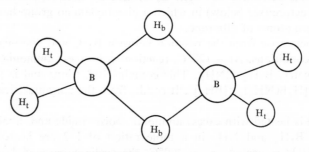

FIG. 1.1. The bridged structure of diborane.

maker and the more recent results of Bartell and Carroll[20] on B_2H_6 and B_2D_6. In the latter, electron diffraction photographs were examined microphotometrically to obtain objective measurements of intensity. There is close agreement between the molecular parameters found from these two studies.

In the solid state between its melting-point, $-165\,°C$, and liquid nitrogen temperature, diborane is believed to be capable of existing in three phases, α, β and ω. The α- and β-forms were first characterized by their X-ray powder patterns[21] and were found to co-exist over a considerable range of temperature. It is therefore likely that the two phases are closely related structurally.

α-Diborane is deposited from diborane vapour at $-268·8\,°C$ and transforms slowly into β-diborane at $c. -213\,°C$. β-Diborane has also been made by deposition from the vapour state at $-196\,°C$ followed by annealing above $-183\,°C$.

TABLE I

Molecular Parameters for B_2H_6 and B_2D_6 from Electron Diffraction Data

	B_2H_6		B_2D_6
	Hedberg and Schomaker[19]	Bartell and Carroll[20]	Bartell and Carroll[20]
B, B (nm)	0·177 ± 0·0013	0·1775 ± 0·0003	0·1771 ± 0·0003
B, H_t(nm)	0·1187 ± 0·003	0·1196 ± 0·0007	0·1198 ± 0·0006
B, H_b(nm)	0·1334 ± 0·0027	0·1339 + 0·0002 − 0·0006	0·1333 + 0·0002 − 0·0004
H_t–B–H_t angle (deg.)	121·5 ± 7		
H_b–B–B angle (deg.)	48.5	48·5 ± 0·15	48·4 ± 0·15
B–B–H_t angle (deg.)		120·6 ± 0·9	119·4 ± 0·9

The X-ray diffraction analysis of β-diborane has been carried out[22] and this shows the crystal is monoclinic with the molecular parameters:

B, B $= 0·1776 ± 0·001$ nm; B, $H_t' = 0·108 ± 0·002$ nm;
B, $H_t'' = 0·110 ± 0·002$ nm; B, $H_b' = 0·123 ± 0·002$ nm;
B, $H_b'' = 0·125 ± 0·002$ nm; H_t'–B–$H_t'' = 124 ± 1°$;
H_b'–B–$H_b'' = 90 ± 1°$.

The B, B distance agrees well with the values determined by electron diffraction. The B, H bond lengths are systematically about 0·01 nm shorter than the electron diffraction values. This shortening appears to be a general phenomenon[23] associated with parameters determined by X-ray crystallography and so the electron diffraction data may still be regarded as reliable.

The α- and β-phases have been differentiated by differences in their polarized infrared spectra.[24] β-Diborane shows a very intense component of absorption on the low-frequency side of the band at 1600 cm^{-1}, but there is no corresponding feature in the spectrum of the α-form. From a detailed study of the intensity of the infrared absorption bands of the α-form as a function of polarizer setting, it has been concluded that α-diborane is monoclinic. Its detailed structure has not yet been published.

The ω-form has been obtained as single crystals, but within hours or a few days these transform spontaneously into polycrystalline masses. Its structure is unknown.

Vibrational Spectra

Raman data[25] on liquid diborane and the infrared spectrum[26] of the gas were at one time regarded as consistent with an ethane-like molecular structure. It was, however, recognized that the infrared spectrum of diborane was much more complex than that of ethane, but this was erroneously attributed to the existence of some low-lying excited electronic levels which could participate in some of the vibrational transitions.

Ethane has D_{3d} or D_{3h} symmetry: the bridge structure of diborane, assuming the boron valencies are tetrahedrally disposed, belongs (like ethylene) to the symmetry group D_{2h}. Bell and Longuet-Higgins[27] have calculated, by normal coordinate analysis, the frequencies of the fundamental vibrations expected for the bridge structure and have found generally a satisfactory agreement with experimental results.

The number of fundamental vibrations for D_{2h} symmetry, their classes and activities are summarized in Table II. There are no degenerate vibrations and none are active both in the Raman and in the infrared spectrum because the molecule has a centre of symmetry. These are two major features in which the bridge structure differs from an ethane structure.

TABLE II

Normal Vibrations of the B_2H_6 Molecule (bridge structure)

(p) = polarized; (dp) = depolarized

Polarized Raman lines correspond with vibrations of the molecule which are totally symmetric: depolarized lines correspond with vibrations that are not totally symmetric.

Class	Activity	Vibration
A_{1g}	R(p)	$\nu_1, \nu_2, \nu_3, \nu_4$
A_{1u}	inactive	ν_5
B_{1g}	R(dp)	ν_6, ν_7
B_{1u}	i.r.	ν_8, ν_9, ν_{10}
B_{2g}	R(dp)	ν_{11}, ν_{12}
B_{2u}	i.r.	ν_{13}, ν_{14}
B_{3g}	R(dp)	ν_{15}
B_{3u}	i.r.	$\nu_{16}, \nu_{17}, \nu_{18}$

TABLE III

Infrared Spectra of Diborane Vapour[29] (frequencies in cm^{-1})

$^{10}B_2H_6$	Assignment	$^{10}B_2D_6$	Assignment
368 s	ν_{10}	262	ν_{10}
829 vw	ν_5	592 vvw	ν_5
977 s	ν_{14}	728 s	ν_{14}
1181 vs	ν_{18}	881 vs	ν_{18}
1298 w	$\nu_{10} + \nu_{12}$	981 ⎫	
1374 w	$\nu_{10} + \nu_{15}$	989 ⎬ w	$\nu_{10} + \nu_{12}$
1606 vvs	ν_{17}	997 ⎭	$\nu_{10} + \nu_{15}$
1869 ms	$\nu_5 + \nu_7$	1205 vvs	ν_{17}
1887 m	ν_{13}	1322 m	$\nu_5 + \nu_{15}$
1999 w	$\nu_9 + \nu_{15}$	1406 w	$\nu_9 + \nu_{15}$
2151 vvw	$\nu_3 + \nu_{14}$	1459 ms	$\nu_5 + \nu_7$
2359 m	$\nu_3 + \nu_{18}$	1491 m	ν_{13}
2412 w	$\nu_4 + \nu_{17}$	1799 s	$\nu_3 + \nu_{18}$
2528 vs	ν_{16}	1857 vs	ν_{16}
2625 vs	ν_8	1999 vs	ν_8
2795 vw	$\nu_3 + \nu_{17}$?	2134 w	$\nu_3 + \nu_{17}$
		2219 w	$\nu_2 + \nu_9$
		2704 m	$\nu_2 + \nu_{17}$
		2759 vw	$\nu_3 + \nu_{16}$

Figure 1.2 shows the approximate nature of the fundamental vibrations, as described by Bell and Longuet-Higgins, assuming that the bending force constants are considerably smaller than the stretching ones.

Several studies of the vibrational spectrum of diborane, including notably that by Price,[28] preceded the definitive work of Lord and Nielson.[29] Their band assignments are generally accepted today. They simplified the interpretation of the spectrum by using

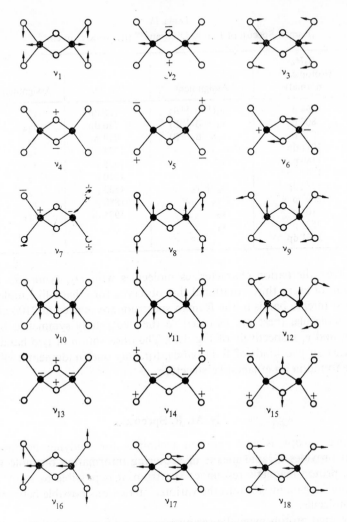

Fɪɢ. 1.2. The normal vibrations of the diborane molecule.

$^{10}B_2H_6$, i.e. diborane containing isotopically pure boron. All infrared spectra prior to this work had been measured on B_2H_6 containing the natural abundances of ^{10}B and ^{11}B. The infrared spectra of $^{10}B_2H_6$ and $^{20}B_2D_6$ are given in Table III, together with assignments of fundamentals and some combination bands. Raman spectra results are given in Table IV. ν_5, ν_7, ν_9 and ν_{12} were not directly observed and their assignments have been made from the frequencies of overtone and combination bands. Lord and Nielson's assignments for $^{10}B_2D_6$ were: ν_5, 592; ν_7, 870; ν_9, 705; ν_{12}, 740 (all in cm^{-1}). Their identification of ν_7 in $^{11}B_2H_6$, 1033 cm^{-1}, is the only assignment which appears to have been challenged. An alternative assignment[24] of ν_7 is at 850 cm^{-1}.

The complexity of the infrared spectrum of diborane arises from the occurrence of eight active fundamental vibrations (cf. five only for ethane). There are striking similarities between the high-resolution spectra of ethylene and diborane, notably in the band contours and in the well-marked alternation of the intensities of the rotational fine structure of

TABLE IV

Raman Spectra of Liquid Diborane[29] (frequencies in cm^{-1})

B_2H_6 (isotopically normal)	Assignment	$^{10}B_2D_6$	Assignment
794 p	$\nu_4(^{11}B-^{11}B)$	726 dp	ν_4
807 p	$\nu_4(^{11}B-^{10}B)$	730 dp	ν_{15}
820 p	$\nu_4(^{10}B-^{10}B)$	929 p	ν_3
1011 dp	ν_{15}	1273 dp	ν_6
1180 p	ν_3	1435 p	$2\nu_{15}$
1747 dp	ν_6	1510 p	ν_2
1788 dp	$\nu_5 + \nu_9$	1833 p	$2\nu_3$
2007 p	$2\nu_{15}$	1880 p	ν_1
2104 p	ν_2	1975 dp	ν_{11}
2524 p	ν_1		
2591 dp	ν_{11}		

some bands. This alternation characterizes molecules with D_{2h} symmetry in contrast to the accentuation of every third rotational line observed for ethane-like molecules.

The two most intense bands in the Raman spectrum are centred at 2104 and 2524 cm^{-1}. For a bridge molecule, these are assigned to the two totally symmetric B–H stretching frequencies (ν_2 and ν_1 respectively of Fig. 1.2). The observation of two bands is consistent with the presence of two kinds of B–H bands, for only one fundamental vibration of this type would be found in an ethane-like structure.

N. M. R. Spectra

N. M. R. spectroscopy is an important technique for studying boranes and related compounds. It provides a rapid means of acquiring information on molecular structure. As well as the proton magnetic resonance spectrum, it is possible to study the resonance due to boron and to gain insight into the various interactions possible between the different nuclei in the molecule.

The first reported proton magnetic resonance spectra were measured on low resolution spectrometers.[30-32] The spectrum at 30 MHz shows (Fig. 1.3a) four main peaks superimposed on weaker resonance bands. The four large peaks are attributed to the terminal protons, the resonance of which is split by coupling between these and a directly bonded ^{11}B nucleus (Fig. 1.3b). For this $I = 3/2$, so there are four $(2I + 1)$ possible orientations. The bridge protons interact with the magnetic fields due to the seven possible orientations ($I_{resultant} = +3, +2, +1, 0, -1, -2$ and -3) of the two ^{11}B atoms equidistant from them. A septet of lines results and their theoretical intensities are in the ratio $1:2:3:4:3:2:1$ (Fig. 1.3d).

Normal diborane consists of the three isotopically different molecules, $^{11}B_2H_6$, $^{11}B^{10}BH_6$ and $^{10}B_2H_6$, which are present, respectively, in approximately 64, 32 and 4% abundance. While the spectrum due to $^{11}B_2H_6$ predominates, there are contributions due to the other species. For example, the ^{10}B nucleus has $I = 3$ and it splits the resonance of the attached terminal protons into seven peaks of equal intensity (Fig. 1.3c).

The interpretation was confirmed by an examination of the proton magnetic resonance of diborane at 30 MHz while simultaneously applying a strong field of 9·6257 MHz, the

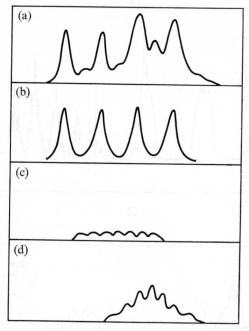

FIG. 1.3. The proton magnetic resonance of diborane at 30 MHz: (a) complete spectrum; (b), (c) and (d) resolved into its components.

resonance frequency of ^{11}B. This technique induces transitions between the various spin states of the perturbing ^{11}B nuclei frequently enough so that the protons whose resonance is being studied are affected by an average value of the spins. The spectrum obtained shows two peaks, the amplitudes of which are in the ratio 4:2, in accordance with the presence of four terminal and two bridge protons.

The ^{11}B resonance spectrum measured at 12·3 MHz has also been interpreted in accordance with a bridged structure for diborane. This consists of a triplet of relative intensities 1:2:1, arising from the splitting of the ^{11}B resonance by the two terminal protons to which it is bound and corresponding with the various combinations of the spin orientations of two protons which are equally probably, viz., $+\frac{1}{2}, +\frac{1}{2}$; $+\frac{1}{2}, -\frac{1}{2}$; $-\frac{1}{2}, +\frac{1}{2}$; $-\frac{1}{2}, -\frac{1}{2}$. Each member of this triplet is split by the two bridge protons into a small triplet of relative intensities, 1:2:1.

The resonances due to the bridge and terminal protons are largely separated at 60 MHz. The spectrum[33] reveals fine structure of the bridge proton septet due to spin–spin coupling between bridge and terminal protons. Each of the seven resonance lines is split thereby into a quintet.

At 100 MHz,[34] there is no overlap of the bridge and terminal proton resonances. Figure 1.4 shows the proton resonance spectrum measured at $-50°$C and the detailed structure of the first two multiplets in the region where the terminal protons resonate. Splitting of each of the four main peaks of the terminal protons into a triplet would be expected at high resolution due to coupling with the bridge protons, but the fine structure is clearly more complex than this. Similarly, the ^{11}B resonance spectrum, a triplet of triplets under low resolution, has a considerably more elaborate structure than this at 19·25 MHz (Fig. 1.5).

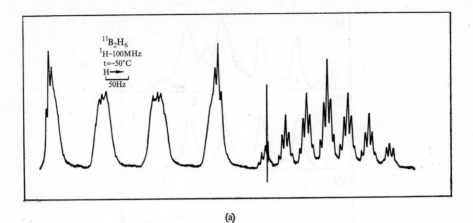

(a)

(b)

FIG. 1.4. The proton magnetic resonance spectrum of $^{11}B_2H_6$ at 100 MHz: (a) complete spectrum; (b) detail of lowfield multiplets (reproduced by permission from Farrar, T. C., Johannsen, R. B., and Coyle, T. D., *J. Chem. Phys.* **49**, 281 (1968)).

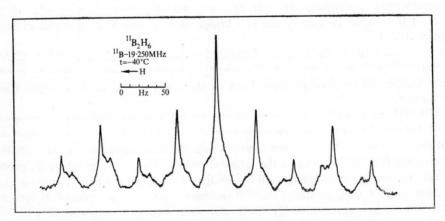

FIG. 1.5. ^{11}B magnetic resonance spectrum of $^{11}B_2H_6$ at 19·25 MHz.

To explain the observed spectra at high resolution, the following interactions must be considered: (a) Coupling between terminal protons and the boron nucleus to which they are bonded, J_{BH_t}; (b) coupling between bridge protons and boron, J_{BH_b}; (c) bridge-proton/terminal-proton coupling, $J_{H_bH_t}$; (d) coupling between the terminal protons and the remote boron atom, J'_{BH_t}; (e) coupling between terminal protons on opposite ends of the molecule, $J_{H_tH_t (cis)}$ and $J_{H_tH_t (trans)}$; (f) geminal coupling between terminal protons, $J_{H_tH_t (gem)}$; (g) boron-boron coupling, J_{BB}; (h) the bridge-proton/terminal-proton chemical shift, $\delta(H_t-H_b)$. Calculated spectra in very good agreement with the observed spectra have been obtained using the parameters for $^{11}B_2H_6$ given in Table V. From comparisons of this kind it has

TABLE V

Coupling Constants and Chemical Shift Data for $^{11}B_2H_6$

	Hz
J_{BB}	∓ 5
J_{BH_b}	$+46 \cdot 2$
J_{BH_t}	$+133$
J_{BH_t}	$+4$
$J_{H_tH_b}$	$7 \cdot 2$
$J_{H_tH_t (trans\ or\ cis)}$	± 14
$J_{H_tH_t (cis\ or\ trans)}$	± 6
$J_{H_tH_t (gem)}$	3

$$\delta(H_t-H_b) = +4 \cdot 50 \text{ ppm}$$

been found that the spectrum of the bridge protons is of first order and the bridge proton resonance is affected only by variation of the parameters $J_{H_bH_t}$ and J_{BH_b}. The model of B_2H_6 with the terminal protons equivalent in pairs appears to be incorrect and these protons must be regarded as magnetically non-equivalent.

Thermochemical Data

The enthalpy of formation of diborane has been calculated indirectly using two different Born–Haber cycles based respectively on the formation of the trimethylamineborane adduct and on the hydrolysis of diborane to orthoboric acid. In the first of these:

$$(CH_3)_3N(g) \quad + \quad 0 \cdot 5B_2H_6 \text{ (g)} \xrightarrow{-135 \cdot 3} (CH_3)_3N \cdot BH_3 \text{ (c)}$$

$$\uparrow^{-24 \cdot 7} \qquad\qquad \uparrow^{\frac{1}{2}\Delta H^0 298 \cdot 15} \qquad \uparrow^{-142 \cdot 4}$$

$$3C \text{ (c)} + \tfrac{1}{2}N_2 \text{ (g)} + 9/2H_2 \text{ (g)} + 3/2H_2 \text{ (g)} + B \text{ (c)} \overline{\qquad\qquad\qquad\qquad}$$

(enthalpies in kJ mol^{-1})

the enthalpies of formation of $(CH_3)_3N$ and $(CH_3)_3N \cdot BH_3$ and the enthalpy of reaction between $(CH_3)_3N$ and B_2H_6 have been determined experimentally.[35] The calculated value of $\Delta H^0_{298 \cdot 15} = 35 \cdot 2$ kJ mol^{-1}.

From a similar cycle and data on the enthalpies of formation and solution of boric acid, the enthalpy of hydrolysis of diborane[36] and the enthalpy of formation of water,

$\Delta H^0_{298.15} = 39 \cdot 8$ kJ mol^{-1}. Using an alternative value $-466 \cdot 6$ kJ mol^{-1} for the enthalpy of solution of orthoboric acid, [37]$\Delta H_{298.15} = 36 \cdot 7$ kJ mol^{-1}.

These values show satisfactory agreement with one another although they are all larger than the value of $31 \cdot 5$ kJ mol^{-1} derived from calorimetric studies on the decomposition of diborane into amorphous boron and hydrogen gas.[38] This value is probably less reliable because of uncertainty in one of the quantities used in its calculation, the enthalpy of formation of amorphous boron.

In view of the dimeric nature of diborane and the considerable interest in the kind of bonding within the molecule, several determinations of the dissociation energy, the energy required to cause symmetrical cleavage into two borane radicals, have been carried out.

The dissociation energy has been calculated from kinetic and equilibrium studies on the system:

$$2BH_3CO \rightleftharpoons B_2H_6 + 2CO$$

Borane carbonyl is a convenient source of BH_3 which combines, as soon as it is formed, to give B_2H_6. Two recent studies have fixed the upper limit for the dissociation energy between 140[39] and 160[40] kJ mol^{-1}. A mass spectrometric investigation[41] of the low-pressure pyrolysis of borane carbonyl has given the value of $150 \pm 12 \cdot 5$ kJ mol^{-1}.

The compounds $BH_3 \cdot PF_3$, $BH_3 \cdot CF_3PF_2$ and $BH_3(CF_3)_2PF$ all decompose to B_2H_6 and the free phosphine ligand. The kinetics of their decomposition have been studied by infrared spectroscopy,[42] exploiting the very intense peaks due to P–F stretching vibrations which occur at different frequencies in the free ligand compared with the borane adduct. Intensity measurements serve as a quantitative determination of the ligand as it is released. From this work, the dissociation energy of diborane was found to be 146 kJ mol^{-1}.

Bonding

The structure of diborane cannot be expressed in terms of conventional covalent bonds because the boron atom has only three electrons in its valence shell and yet is able to form bonds with four hydrogen atoms.

Several theories have been proposed which attempt to describe the bonds in molecules where the number of valence electrons is insufficient.

According to resonance theory, bonding in B_2H_6 involves contributions from resonance structures like

Each of the bonds shown consists of a shared pair of electrons. This representation of the structure is not satisfactory since it implies that the two parts of the molecule are held together largely by resonance energy. It also fails to give real insight into the nature of the bonds which form the bridge.

Pitzer[43] proposed the bridge should be regarded as a protonated double bond

$$\begin{array}{c} H \quad\quad H^+ \quad\quad H \\ \diagdown \quad\quad | \quad\quad \diagup \\ \bar{B} =\!\!=\!\!= \bar{B} \\ \diagup \quad\quad | \quad\quad \diagdown \\ H \quad\quad H^+ \quad\quad H \end{array}$$

in which the boron atoms are joined by a σ- and a π-bond. The protons are embedded in the π-electron cloud above and below the plane of the rest of the molecule. This model accounts for the known resistance to internal rotation within diborane and the resemblance in spectroscopic properties of diborane to ethylene rather than ethane. On the other hand, the observed B,B distance is much greater than that expected for a double bond and there is no chemical evidence to support the suggestion that the bridge hydrogens are acidic. Indeed, chemical shift data indicate quite the contrary: the bridge hydrogens resonate at higher field than the terminal protons. They are therefore more shielded than the latter and must be thought of as hydride-like in character.

Electron deficiency in B_2H_6 is a consequence of there being more atomic orbitals available for bonding than there are valence electrons. For normal covalent bond formation, two orbitals, one from each of the combining atoms, overlap to give a bonding and an antibonding orbital. When two electrons are available, the bonding orbital is filled and this constitutes the covalent bond. In the case of B_2H_6, it has been proposed that the bonding of the bridge hydrogens involves two 'three-centre' bonds. A three-centre bond results from the interaction of three orbitals, each on a different atom, to form one bonding, one antibonding and one essentially non-bonding orbital. Two electrons suffice to fill the bonding orbital which encompasses the three atoms.

A simple description of the structure of diborane is as follows: Each B atom is sp^3 hybridized. The terminal B–H bonds are formed by overlap of sp^3 hybrid orbitals with $1s$ orbitals on the hydrogens and they are of the normal electron-pair type. The very small difference in the electronegativities of B and H mean that these bonds are effectively non-polar. Each B–H–B bridge bond is a three-centre bond formed from a $1s$ orbital on hydrogen and a sp^3 hybrid orbital from each boron. The hydrogen atom contributes one electron and each boron formally gives half an electron to this bond. It is again generally assumed to be non-polar, although a theoretical LCAO treatment[44] of the orbitals in diborane, in which the bonding was considered as a four-centre four-electron problem involving the bridge hydrogen orbitals and the boron orbitals directed towards them, indicated that the bridge protons carry about a charge of about $-0.2e$. The localized three-centre bonding orbitals are illustrated in Fig. 1.6. It is important to note that only two electrons are needed for a three-centre bond. Even if more electrons were available, they would enter a non-bonding orbital and so would not increase the bond strength.

There has been a recent revival of interest in structures where there is some kind of direct bond between the boron atoms. For example, a comparison[45] between calculated and observed frequencies for the normal vibrations of diborane has shown that the best fit is obtained using a potential function with contributing terms involving not only the bridge (B–H_b) and terminal (B–H_t) bond stretching force constants, k_b and k_t, but also k_l, the stretching constant for a direct B–B bond. The calculated force constants k_b, k_t and k_l are respectively 176·8, 354·1 and 272 Nm^{-1}. Values for k_b and k_t are consistent with a considerably weaker bridge than terminal bond and that for k_l with the existence of a B–B bond.

13

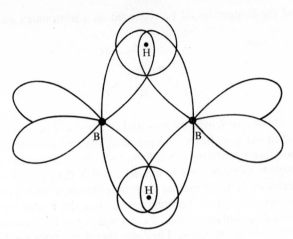

FIG. 1.6. Localized three-centre bonding orbitals in diborane.

To describe the bonding in B_2H_6, it is pertinent to remember that the H_t–B–H_t and B–B–H_t bond angles are both in the region of 120°. These would be consistent with sp^2 hybrid orbitals on each boron atom, forming two B–H bonds with the terminal hydrogens and one bond with the other boron. To facilitate the theoretical treatment,[46] the two bridge hydrogens are regarded as a united atom which has a pseudo σ-orbital, ψ_0, and a pseudo π-orbital, ψ_1, formed by linear combinations of the two atomic $1s$ orbitals.

ψ_0 interacts with the B–B bonding orbitals to form σ-bonding and σ-antibonding molecular orbitals. A π-bonding orbital is also formed from ψ_1 and the B–B orbitals. There are four electrons in the bridge bonds: two come from the bridge hydrogens and two from the borons. The M. O. of lowest energy is the σ-bonding orbital which contains two electrons. The σ-antibonding is considered to have a higher energy than the π-bonding orbital and so the remaining two electrons occupy the latter. The bonding situation is therefore very similar to that in ethane and is not unlike Pitzer's concept of a protonated double bond.

References

1. DAVY, H., *Phil. Trans.* **99**, 85 (1809).
2. STOCK, A., *Hydrides of Boron and Silicon*, Cornell Univ. Press, Ithaca, New York (1933).
3. SCHLESINGER, H. I., and BURG, A. B., *J. Amer. Chem. Soc.* **53**, 4321 (1931).
4. FINHOLT, A. E., BOND, A. C., Jr., and SCHLESINGER, H. I., *J. Amer. Chem. Soc.* **69**, 1199 (1947).
5. *Inorganic Syntheses*, vol. XI, p. 15.
6. DUKE, B. J., GILBERT, J. R., and READ, I. A., *J. Chem. Soc.* 540 (1964).
7. FREEGUARD, G. F., and LONG, L. H., *Chem. and Ind.* 471 (1965).
8. FORD, T. A., KALB, G. H., McCLELLAND, A. L., and MUETTERTIES, E. L., *Inorg. Chem.* **3**, 1032 (1964).
9. BARTON, L., and NICHOLLS, D., *Proc. Chem. Soc.* 242 (1964).
10. CUELLERON, J., and BOUIX, J., *Bull. Soc. Chim.* 2945 (1967).
11. EASTHAM, J. F., *J. Amer. Chem. Soc.* **89**, 2237 (1967).
12. KELLER, P. C., *J. Amer. Chem. Soc.* **91**, 1231 (1969).
13. McACHRAN, G. E., and SHORE, S. G., *Inorg. Chem.* **4**, 125 (1965).
14. BELL, R. P., and EMELEUS, H. J., *Quart. Rev.* **2**, 132 (1948).
15. MARK, H., and POHLAND, E., *Z. Kryst.* **62**, 103 (1925).
16. BAUER, S. H., *J. Amer. Chem. Soc.* **59**, 1096 (1937).
17. BAUER, S. H., *Chem. Rev.* **35**, 180 (1942).

18. DIATKINA, M. E., and SIRKIN, J. R., *Compt. Rend. Acad. Sci. U.R.S.S.* **35**, 180 (1942).
19. HEDBERG, K., and SCHOMAKER, V., *J. Amer. Chem. Soc.* **73**, 1482 (1951).
20. BARTELL, L. S., and CARROLL, B. L., *J. Chem. Phys.* **42**, 1135 (1965).
21. BOLZ, L. H., MAUER, F. A., and PEISER, H. S., *J. Chem. Phys.* **31**, 1005 (1959).
22. SMITH, H. W., and LIPSCOMB, W. N., *J. Chem. Phys.* **43**, 1060 (1965).
23. LIPSCOMB, W. N., *J. Chem. Phys.* **22**, 985 (1954).
24. FREUND, I., and HALFORD, R. S., *J. Chem. Phys.* **43**, 3795 (1965).
25. ANDERSON, T. F., and BURG, A. B., *J. Chem. Phys.* **6**, 586 (1938).
26. STITT, F., *J. Chem. Phys.* **9**, 780 (1941).
27. BELL, R. P., and LONGUET-HIGGINS, H. C., *Proc. Roy. Soc.* A **138**, 357 (1945).
28. PRICE, W. C., *J. Chem. Phys.* **16**, 894 (1948).
29. LORD, R. C., and NIELSON, E., *J. Chem. Phys.* **19**, 1 (1951).
30. KELLY, J., Jr., RAY, J., and OGG, R. A., Jr., *Phys. Rev.* **94**, 767 A (1954).
31. OGG, R. A., Jr., *J. Chem. Phys.* **22**, 1933 (1954).
32. SHOOLERY, J. N., *Disc. Faraday Soc.* **19**, 215 (1955).
33. GAINES, D. F., SCHAEFFER, R., and TEBBE, F., *J. Phys. Chem.* **67**, 1937 (1963).
34. FARRAR, T. C., JOHANNSEN, R. B., and COYLE, T. D., *J. Chem. Phys.* **49**, 281 (1968).
35. GOOD, W. D., and MÅNSSON, M., *J. Phys. Chem.* **70**, 97 (1966).
36. GUNN, S. R., and GREEN, L. G., *J. Phys. Chem.* **64**, 61 (1960).
37. PROSEN, E. J., JOHNSON, W. H., and PERGIEL, F. Y., *J. Res. Natl. Bur. Std.* **62**, 43 (1959).
38. GUNN, S. R., and GREEN, L. G., *J. Phys. Chem.* **65**, 779 (1961).
39. GROTEWOLD, J., LISSI, E. A., and VILLA, A. E., *J. Chem. Soc.* A 1038 (1966).
40. GARABEDIAN, H. E., and BENSON, S. W., *J. Amer. Chem. Soc.* **86**, 176 (1964).
41. FEHLNER, T. P., and MAPPES, G. W., *J. Phys. Chem.* **73**, 873 (1969).
42. BURG, A. B., and FU, Y.-C., *J. Amer. Chem. Soc.* **88**, 1147 (1966).
43. PITZER, K., *J. Amer. Chem. Soc.* **67**, 1126 (1945).
44. HAMILTON, W. C., *Proc. Roy. Soc.* A **235**, 395 (1956).
45. OGAWA, T., and MIYAZAWA, T., *Spectrochim. Acta* **20**, 557 (1964).
46. OGAWA, T., and HIROTA, K., *Bull. Chem. Soc. Jap.* **40**, 237 (1967).

CHAPTER 2

BORAZINE

Introduction

Borazine, $B_3N_3H_6$, is of special interest because of its molecular structure, a six-membered ring of alternate boron and nitrogen atoms. It is isoelectronic with benzene and in some of its physical properties shows quite striking similarities with this (Table I). Indeed, the earlier name borazole was proposed because of the resemblance to benzene ('benzol') and it has been frequently referred to as 'inorganic benzene'. The systematic name is s-triazatriborine, but the compound is still generally known as borazine.

TABLE I

Comparison of the Physical Properties of Borazine and Benzene

Compound	Molecular weight	M.p. °C	B.p. °C	$\Delta H_{vap.}$ (kJ mol^{-1})	Trouton's constant
Borazine	80·5	−58	54·5	29·3	21·4
Benzene	78	+ 6	80	31	21·1

The heterocyclic nature of the borazine ring means that there are significant differences between the chemistry of benzene and borazine. It is one of a number of six-membered heterocyclic compounds including γ-SO_3, the trimetaphosphate ion, $P_3O_9{}^{3-}$, and trimeric chlorophosphazene, $(PNCl_2)_3$. The detailed study of such ring compounds is comparatively new in inorganic chemistry in contrast to the many years for which organic heterocyclics have been known and examined.

Preparation

Borazine (I) was originally prepared by Stock and Pohland[1] in 1926 by the reaction of ammonia with diborane. The adduct $BH_3 \cdot NH_3$ is first formed and then decomposed by heating in a closed tube at 200°C. Cyclization occurs to give borazine, but yields are low because of the simultaneous formation of solid polymeric by-products. Much attention has been given to devising more convenient syntheses because of the difficulties in preparing and handling boranes.

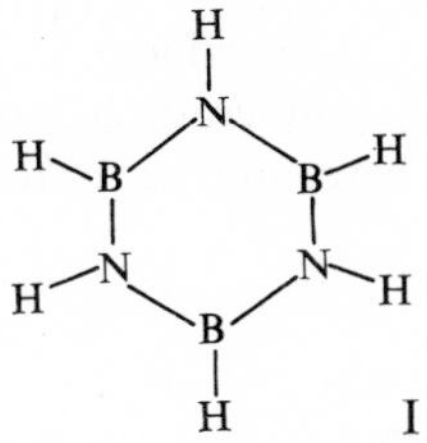

I

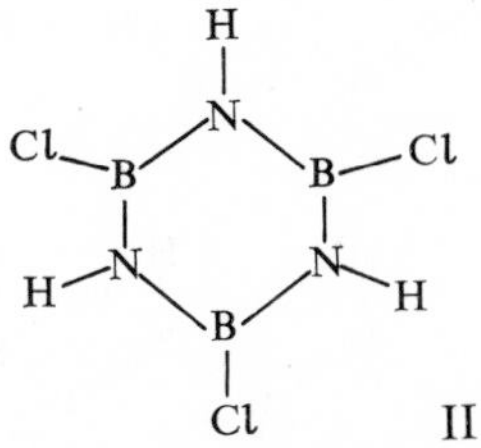

II

B-Trichloroborazine (II), formed by the reaction between boron trichloride and ammonium chloride, can serve as an intermediate in the preparation of borazine.[2] Small quantities are conveniently made by heating dry, powdered ammonium chloride in one end of a Pyrex combustion tube at 170°C while a slow stream of boron trichloride vapour and nitrogen is passed over the powder. $B_3N_3H_3Cl_3$ sublimes as it is formed and condenses as needle-like crystals on the cool parts of the tube. Yields (based on the amount of ammonium chloride) of up to 35% have been reported. Non-volatile solids are formed as by-products. Similar yields are obtained if the reaction is carried out in the presence of refluxing chlorobenzene. Impure B-trichloroborazine is obtained by evaporation of the chlorobenzene after filtration of the cooled reaction product. It is readily purified by vacuum sublimation.

B-Trichloroborazine is reduced to borazine by metal tetrahydroborates.[3]

$$3MBH_4 + B_3N_3H_3Cl_3 \rightarrow B_3N_3H_6 + 1{\cdot}5B_2H_6 + 3MCl$$

As the equation shows, 1·5 moles of diborane are liberated for every mole of $B_3N_3H_3Cl_3$ used. This necessitates special precautions and the reaction is difficult to exploit on a large scale unless immediate use can be made of the diborane or suitable storage facilities are available. An alternative procedure is to prevent the evolution of diborane by the addition of a tertiary aliphatic amine of relatively high boiling-point, such as tri-n-butylamine.[4] This reacts rapidly and completely with diborane in situ to give the solid amine/borane adduct.

This reaction is the basis of a convenient laboratory preparative method for borazine.[5] Sodium tetrahydroborate, dissolved in diglyme (the dimethyl ether of diethyleneglycol), is used as the reducing agent. Borazine is distilled in vacuo and purified by fractionation. A yield of 46% of pure borazine, a colourless liquid (b.p. = 54·5°C), has been reported.

Borazine decomposes slowly on storage, even at −80°C, and small quantities of a white solid are deposited over a period of several days. At room temperature, decomposition

17

is much more rapid and hydrogen, diborane and other volatile products are formed. Explosions of highly purified samples sealed in small ampoules have occurred when these were stored for several weeks in normal day-to-day illumination. Samples can be safely stored in the dark for several months.

Reactions

When borazine is pyrolysed[6] at temperatures above 340°C, the products include $B_6N_6H_{10}$ (III) and $B_5N_5H_8$ (IV). These are the boron–nitrogen analogues of diphenyl

and naphthalene respectively. B-2,4-diaminoborazine is also formed, presumably after a breakdown of some of the borazine molecules. The thermal decomposition of borazine is not completely analogous to that of benzene, which gives biphenyl, terphenyls and triphenylene but no naphthalene. The decomposition of liquid borazine at ambient temperature[7] produces $B_6N_6H_{10}$, $B_5N_8H_{11}$ (the B-triamino derivative of $B_5N_5H_8$) and B-2,4-diaminoborazine.

Borazine forms an adduct[8] with methanol in which borazine : methanol = 1:3. This pyrolyses with the elimination of hydrogen to give B-trimethoxyborazine.

Borazine reacts with phenylmagnesium bromide to give B-arylated compounds.[9] The main products are B-mono- and B-diphenylborazine together with a small amount of B-triphenylborazine. Borazine reacts with a deficiency of methylmagnesium iodide in ether solution to produce a mixture of 2-methyl, 2,4-dimethyl, and 2,4,6-trimethyl-B-borazines.[10]

In general, the borazine ring undergoes substitution reactions primarily at the boron atoms and much less commonly at the nitrogens. N-methylborazines can be formed,[10] however, when ammonium chloride and N-methylaminoborane trimer $(H_2BNHCH_3)_3$ are heated together at 125–150°C. A mixture of mono-, di- and tri-N-methylborazines and borazine itself is obtained.

Although it was once believed that one mole of borazine forms an adduct with two moles of bromine and that this decomposes to B-2,4-dibromoborazine as the main product,[11] these reactions have not been repeated.[12] It appears that borazine reacts with an excess of bromine to give an orange solid, B-hexabromocyclotriborazane, $B_3N_3H_6Br_6$, and this decomposes on heating to give B-tribromoborazine and HBr.

Borazine undergoes a strongly exothermic reaction with aniline[13] to produce triamino-borine (V).

V

Halogen-substitued borazines possess reactive B–H, B–halogen and N–H bonds and are useful intermediates in synthetic work. The reaction of borazine with boron trichloride[14] gives B-2-chloroborazine and B-2,4-dichloroborazine. This reaction is very slow and only small yields of the halogen-substituted borazines are obtained. Better yields are obtained by reaction between borazine and mercury(II) chloride in ether or n-pentane at room temperature.[14]

Some other borazine derivatives have been made by photochemical reactions[15] in which the borazine molecule is activated by irradiation with ultraviolet light. Mixtures of borazine and oxygen at low pressure give B-monohydroxoborazine, $B_3N_3H_5OH$, and diborazinyl ether, $(B_3N_3H_5)_2O$.

The borazine analogue of aniline, B-monoaminoborazine, $B_3N_4H_7$, has been made by the photochemical reaction of borazine with ammonia.

Structure

X-ray crystallography has shown[17] that the boron and nitrogen atoms in borazine are arranged alternately in a planar, hexagonal ring, in which the B,N bond length is 0.147 ± 0.007 nm. Electron diffraction photographs of borazine and benzene are very similar and confirm that borazine has a benzene-like configuration. The B,N bond length determined by electron diffraction[18] is 0.144 ± 0.002 nm.

A planar borazine ring should have a zero dipole moment. However, the values of 0.67 and 0.50 D† have been reported for borazine in the vapour phase[19] and in benzene solution[20] respectively. It was recognized at the time that the vapour phase measurements were made that they could be unreliable because of the instability of borazine. The dipole moment in benzene solution was calculated from dielectric constant measurements by standard methods. The observed moment could arise from a permanent deviation of the ring from co-planarity, but this seems most unlikely in view of the X-ray and electron diffraction results. It has been proposed alternatively[20] that low-frequency molecular vibrations occur which lead to a separation of charge and the establishment of a temporary moment. This could result in a large atomic polarization even though the molecule possesses no permanent polarity. In this context, it is interesting to note that some doubt has been expressed[21] concerning the apparent dipole moments of certain symmetrically-substituted derivatives of benzene. Further work is needed to resolve these discrepancies. The evidence

† One Debye unit (D) is equivalent, in S.I. units, to 3.3356×10^{-30} Cm. In view of the smallness of the conversion factor, it is convenient to retain the Debye as the unit of dipole moment.

19

for a permanent dipole moment for borazine is not completely conclusive and the planarity of the molecule is therefore not seriously in doubt.

The ultraviolet absorption spectra of borazine and benzene in n-heptane show[22] a strong resemblance to each other. The four diffuse maxima observed for borazine at 199·5, 196·2, 192·8 and 189·5 nm respectively are analogous to the series of peaks for benzene beginning at 208 nm. Strong absorption in borazine solution rises to a maximum below 172 nm and this corresponds with a strong benzene peak with its maximum at 183 nm. It is probable that the transitions responsible for these absorption bands occur between electronic energy levels which are very similar in both molecules. Borazine thus appears to show the same kind of electron delocalization so characteristic of benzene.

Vibrational Spectrum

The vibrational spectrum of borazine has been interpreted according to the expectations for a planar molecule of D_{3h} symmetry. The fundamental vibrations and their classes, activities and properties are given in Table II.

TABLE II

Fundamental Vibrations of Borazine

Class	Activity	Properties	Vibration
A_1'	R(p)	In phase In phase	ν_1 N–H sym. stretch ν_2 B–H sym. stretch ν_3 B–N ring sym. stretch ν_4 B–N ring bending
A_2'	inactive	In plane In phase	ν_5 B–N ring asym. stretch ν_6 B–H bending ν_7 N–H bending
A_2''	i.r.	Out of plane In phase	ν_8 B–H bending ν_9 N–H bending ν_{10} B–N torsion
E'	R(p) and i.r.	In plane Out of phase	ν_{11} N–H asym. stretch ν_{12} B–H asym. stretch ν_{13} B–N asym. ring bending ν_{14} B–N asym. ring bending ν_{15} B–H bending ν_{16} N–H bending ν_{17} B–N ring bending
E''	R(dp)	Out of plane Out of phase	ν_{18} B–H bending ν_{19} N–H bending ν_{20} B–N bending

Crawford and Edsall[23] carried out the first study of the Raman spectrum of liquid borazine and of the infrared spectrum of its vapour. Their vibrational assignment was based on the observed intensities of the infrared bands and the depolarization ratios of the Raman lines and on a normal coordinate analysis. The resolving power of the infrared spectrometer used was not high and some features of the spectrum were not observed. Price and co-workers[24] measured the infrared spectrum between 400 and 10,000 cm^{-1} with higher resolution and assigned almost all of the observed bands in terms of the fundamentals proposed by Crawford and Edsall. Watanabe and co-workers[25] synthesized N-trideuteroborazine and compared its infrared spectrum with that of borazine. The detailed analysis of the band shapes, frequencies and intensities in the two spectra made possible, for the first time, an unequivocal assignment of the fundamentals. For example, the band at 3490 cm^{-1} (assigned to an N–H stretching mode) disappears on deuteration and is replaced by a band at 2607 cm^{-1}, evidently due to the N–D stretching mode. The assignment of the band at 2530 cm^{-1} in borazine to the B–H stretching frequency is confirmed by the observation of the same band at 2527 cm^{-1} after isotopic substitution. This vibrational mode should be very little affected by the replacement of hydrogen bound to nitrogen by deuterium.

The technique of isotopic labelling of the borazine molecule has been extended[26] with the synthesis of B-deuteroborazine and of borazine enriched with respect to one or other of the stable boron isotopes, ^{10}B and ^{11}B. The complete assignment by Niedenzu et al. for borazine, N-deuteroborazine and B-deuteroborazine is given in Table III. The effects

TABLE III

Assignments of Fundamental Vibrations (cm^{-1}) in Borazine

	(–BD–NH)$_3$	(–BH–NH)$_3$	(–BH–ND)$_3$
ν_1	3452	3452	2579
ν_2	1893	2535	2521
ν_3	903	940	940
ν_4	852	852	824
ν_8	808	917·5	900
ν_9	716	719	546
ν_{10}	326·5	394	383·5
ν_{11}	3485	3486	2594
ν_{12}	1897	2520	2519
ν_{13}	1440	1465	1438
ν_{14}	1328	1406	1289
ν_{15}	1022	1096	1071
ν_{16}	813	990	786
ν_{17}	509	518	507
ν_{18}	725	968	960
ν_{19}	788	798	550
ν_{20}	262	288	283

of substitution of deuterium for hydrogen are clearly seen in the frequency shifts. In addition to the usual spectroscopic selection rule, the assignment of fundamentals is assisted by a study of the contours of the infrared bands observed in the spectra of gaseous samples. Thus borazine may be classified as an oblate symmetric top molecule, the moments of inertia of which are related by $I_A = 2I_B = 2I_C$. For such a molecule, the vibrations of class A_2'' should lead to the absorption bands with a PQR structure, having a prominent central Q branch. In the spectrum of borazine, three bands are observed which have these features. They are at 394, 719 and 917·5 cm^{-1} and are therefore assigned to ν_{10}, ν_9 and ν_8 respectively. An independent study[27] on deuterated borazines has confirmed these assignments.

N.M.R. Spectra

The proton magnetic resonance spectrum of borazine measured[28,29] at 40 MHz is shown in Fig. 2.1. The chemical shift (in ppm), relative to cyclohexane as internal standard, is given along the abscissa. The spectrum consists of a triplet centred at −4·0 and less intense, broad peaks at 2·1, −1·3 and 8·2.

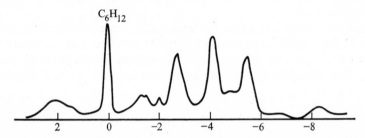

C_6H_{12}

2	0	−2	−4	−6	−8

FIG. 2.1. The proton magnetic resonance spectrum of borazine at 40 MHz.

The triplet arises from the splitting of the proton resonance in NH by coupling with the ^{14}N nucleus. For this, $I = 1$ and there are three orientations ($+1, 0$ and -1) which this can take up with respect to the proton spin. The coupling constant, $J_{NH} = 56$ Hz. This triplet does not show the expected 1:1:1 ratio of intensities, the central line being stronger than the two outer ones which are both of the same intensity. This is due to spin-lattice relaxation (i.e. change of state of spin) of the electric quadrupole moment of the ^{14}N nuclei because of fluctuating electric field gradients arising from the molecular motions of the solvent and solute. This relaxation results in a broadening of the resonance lines, that of the outer components of the proton triplet being greater than that of the central one.[30]

The proton signal of ^{11}BH will split into four equally spaced peaks due to coupling with ^{11}B ($I = 3/2$). Probably the three peaks at $+2\cdot1$, $-1\cdot3$ and $-8\cdot2$ are due to this kind of proton. Then the fourth peak would be expected at $-4\cdot8$. It appears as a small hump between two peaks of the ^{14}NH triplet and is partly masked by it. The spin coupling constant, $J_{11BH} = 138$ Hz. This assignment is consistent with the ^{11}B nuclear magnetic resonance of borazine which shows two lines due to splitting by the proton spin. The separation is 140 Hz, in close agreement with the coupling constant obtained from the proton magnetic resonance spectrum.

Peaks at $+1\cdot4$ and $-1\cdot4$ have been assigned as spinning side bands because they vary in position with the rate of spinning of the sample. Other weak peaks could be due to decomposition products or to the proton resonances of ^{10}BH. For ^{10}B, $I = 3$, and seven proton resonance peaks would be expected. Some of these would probably lie within the region of the strong ^{14}NH resonance and therefore be difficult to detect. Any further assignments in the spectrum measured at 40 MHz would be very speculative.

Bonding

The bonding in borazine has been described in terms of resonance between the single-bonded triaminotriborine ring (I) and the Kekulé structures (Ia) and (Ib) in which double

$$
\begin{array}{c}
\text{H} \\
| \\
\text{N} \\
\text{H--B} \quad \text{B--H} \\
| \qquad | \\
\text{H--N} \quad \text{N--H} \\
\text{B} \\
| \\
\text{H}
\end{array}
$$

Ia

$$
\begin{array}{c}
\text{H} \\
| \\
\text{N} \\
\text{H--B} \quad \text{B--H} \\
\text{H--N} \quad \text{N--H} \\
\text{B} \\
| \\
\text{H}
\end{array}
$$

Ib

bonds are present due to the participation of the lone pairs on nitrogen in the boron–nitrogen bonds. This results in a formal negative charge on the boron and a formal positive charge on the nitrogen atom. This resonance hybrid description is based on the similarities which exist between borazine and benzene. For example, the ring is a planar hexagon and the B,N bond length of 0·144 nm is considerably shorter than the single bond lengths of 0·160 nm in $H_3N \cdot BF_3$ and of 0·162 nm in $(CH_3)_3N \cdot BH_3$. In fact, the B,N bond length is very close to the value in benzene for C,C = 0·142 nm.

The molecular orbital description of borazine takes into consideration the formation of σ- and π-bonds in the ring. Calculations using the extended Hückel theory[31] have shown that the occupied molecular orbitals of highest energy are of the σ-type. Differences in the chemical behaviour of borazine and benzene, such as the greater ease of effecting addition reactions, have been attributed to the greater accessibility of σ-electrons in borazine. π-bonding requires the transfer of electronic charge from nitrogen to boron, but σ-bonding results in a transfer of greater magnitude in the reverse direction. In the resultant charge distribution, the nitrogen atoms carry a formal negative charge and the borons a formal positive charge. This is opposite to the distribution which results from the valence bond description in terms of resonating structures. Nevertheless, it is more in accord with the chemical reactions of borazine and the relative electronegativities of boron and nitrogen which should result in a fairly polar σ-bond between the two atoms in the direction $\overset{\delta+ \quad \delta-}{B-N}$.

A number of other theoretical accounts of the bonding within borazine have been published recently.[32-37] Satisfactory agreement between the energies of electronic transitions calculated theoretically and observed in the ultraviolet spectrum of borazine is found provided the non-uniform distribution of σ-electron charge around the ring is taken into account as well as the π-electron distribution. It has been estimated that about 0·5 of a π-electron is donated to each boron leaving about 1·5 of π-electron density associated with each nitrogen. This transfer is more than offset by the transfer of electronic charge from B to N when the σ-orbitals are formed.

The borazine ring shows[38] the same kind of diamagnetic anisotropy as benzene and graphite. This anisotropy expresses the difference between values of the diamagnetic susceptibility measured along axes parallel to and perpendicular to the molecular plane and is associated with the freedom of movement of π-electrons around the ring. Some doubt exists over the exact magnitude of the diamagnetic anisotropy of borazine. It appears to be generally accepted that it does show some but less than that of benzene. Hence there is some experimental evidence for the mobility of π-electrons within the molecule.

If a criterion of aromaticity is taken to be the extent of delocalization of electrons in π-orbitals, it appears that the 'aromatic' nature of borazine is significantly less than that of benzene. Another feature of aromatic systems is the transmission of electronic effects far from the site of substitution. An experimental study of these is complicated by the polarity of the boron–nitrogen bonds which makes it difficult to prepare all possible isomers. In any case, this polarity would be expected to militate against an effective transmission of electronic charges around the ring. Therefore, one should not try to over-emphasize the similarities between benzene and borazine and it would be more profitable to compare borazine with other inorganic heterocyclic compounds.

References

1. STOCK, A., and POHLAND, E., *Chem. Ber.* **59**, 2215 (1926).
2. BROWN, C. A., and LAUBENGAYER, A. W., *J. Amer. Chem. Soc.* **77**, 3699 (1955).
3. SCHAEFFER, R., STEINDLER, M., HOHNSTEDT, L., SMITH, H. S., EDDY, L. B., and SCHLESINGER, H. I., *J. Amer. Chem. Soc.* **77**, 3303 (1959).
4. DAHL, G. H., and SCHAEFFER, R., *J. Inorg. Nucl. Chem.* **12**, 380 (1960).
5. LAUBENGAYER, A. W., MOEWS, P. C., and PORTER, R. F., *J. Amer. Chem. Soc.* **83**, 1337 (1961).
6. MAMANTOV, G., and MARGRAVE, J. L., *J. Inorg. Nucl. Chem.* **20**, 348 (1961).
7. HAWORTH, D. T., and HOHNSTEDT, L. F., *J. Amer. Chem. Soc.* **81**, 842 (1959).
8. MOEWS, P. C., and LAUBENGAYER, A. W., *Inorg. Chem.* **2**, 1072 (1963).
9. BEACHLEY, O. T., Jr., *Inorg. Chem.* **8**, 981 (1969).
10. WIBERG, E., and BOLZ, A., *Chem. Ber.* **73**, 209 (1940).
11. RILEY, R. F., and SCHACK, C. J., *Inorg. Chem.* **3**, 1651 (1964).
12. KREUTZBERGER, A., and FERRIS, F. C., *J. Org. Chem.* **30**, 360 (1965).
13. SCHAEFFER, G. W., SCHAEFFER, R., and SCHLESINGER, H. I., *J. Amer. Chem. Soc.* **73**, 1612 (1951).
14. MARUCA, R., BEACHLEY, O. T., Jr., and LAUBENGAYER, A. W., *Inorg. Chem.* **6**, 575 (1967).
15. LEE, G. H., and PORTER, R. F., *Inorg. Chem.* **6**, 648 (1967).
16. STOCK, A., and WIERL, R., *Z. Anorg. Chem.* **208**, 228 (1931).
17. BAUER, S. H., *J. Amer. Chem. Soc.* **60**, 524 (1938).
18. RAMASWAMY, K., *Proc. Ind. Acad. Sci.* A **2**, 364, 630 (1935).
19. WATANABE, H., and KUBO, M., *J. Amer. Chem. Soc.* **82**, 2428 (1960).
20. DiCARLO, E. N., and SMYTHE, C. P., *J. Amer. Chem. Soc.* **84**, 1128 (1962).
21. PLATT, J. R., KLEVENS, H. B., and SCHAEFFER, G. W., *J. Chem. Phys.* **15**, 598 (1947).
22. CRAWFORD, B. L., and EDSALL, J. T., *J. Chem. Phys.* **7**, 223 (1939).
23. PRICE, W. C., FRASER, R. D. B., ROBINSON, T. S., and LONGUET-HIGGINS, H. C., *Disc. Faraday Soc.* **9**, 131 (1950).
24. WATANABE, H., TOTANI, T., NAKAGAWA, T., and KUBO, M., *Spectrochim. Acta* **16**, 1076 (1960).
25. NIEDENZU, K., SAWODNY, W., WATANABE, H., DAWSON, J. W., TOTANI, T., and WEBER, W., *Inorg. Chem.* **6**, 1453 (1967).
26. KARTHA, V. B., KRISHNAMACHARI, S. L. N., and SUBRAMANIAM, C. R., *J. Mol. Spectry.* **23**, 149 (1967).
27. ITO, K., WATANABE, H., and KUBO, M., *J. Chem. Phys.* **32**, 947 (1959).
28. ITO, K., WATANABE, H., and KUBO, M., *Bull. Chem. Soc. Jap.* **33**, 1588, (1960).

29. POPLE, J. A., *Mol. Phys.* **1**, 168 (1958).
30. HOFFMANN, R., *J. Chem. Phys.* **40**, 2474 (1964).
31. ROOTHAAN, C. C. J., and MULLIKEN, R. S., *J. Chem. Phys.* **16**, 118 (1948).
32. DAVIES, D. W., *Trans. Faraday Soc.* **56**, 1713 (1960).
33. CHALVET, O., DAUDEL, R., and KAUFMAN, J. J., *J. Amer. Chem. Soc.* **87**, 399 (1965).
34. PERKINS, P. G., and WALL, D. H., *J. Chem. Soc.* A **235** (1966).
35. BOYD, R. J., LO, D. H., and WHITEHEAD, M. A., *Chem. Phys. Lett.* **2**, 227 (1968).
36. DAVIES, D. W., *Trans. Faraday Soc.* **64**, 2881 (1968).
37. WATANABE, H., ITO, K., and KUBO, M., *J. Amer. Chem. Soc.* **82**, 3494 (1960).
38. MUSZKAT, K. A., *J. Amer. Chem. Soc.* **86**, 1250 (1964).

CHAPTER 3

ALUMINIUM CHLORIDE

Introduction

Bridged dimeric molecules are formed by the compounds of other Group III elements beside boron. For example, the lower aluminium trialkyls, such as trimethyl- and triethyl-aluminium, are dimeric in benzene solution and in the vapour state and the same property extends to gallium in trivinylgallium. Stereochemically similar molecules exist for the trichloride, tribromide and triiodide of aluminium. These halides are not necessarily electron-deficient in the same sense as diborane owing to the replacement of hydrogen by halogen, which can, unlike the former element, contribute more than one valence-shell electron in bond formation.

The dimeric molecule is not the only form of aluminium chloride which is known. At high temperatures in the vapour state, dissociation to the monomer, $AlCl_3$, occurs. At low temperatures, when the vapour or liquid is solidified, a major stereochemical change takes place resulting in a predominantly ionic layer lattice. The properties of this compound can be most satisfactorily understood in relation to these various structural forms.

Aluminium chloride has a special importance in organic chemistry because of its well-known catalytic activity in effecting Friedel–Crafts substitution reactions. This depends on the ability of the monomeric molecule to act as a Lewis acid by receipt of an electron pair from a donor atom or molecule, the aluminium atom thereby completing an outermost shell of eight electrons.

Preparation

Anhydrous aluminium chloride is prepared by heating the metal in hydrogen chloride or chlorine vapour. For example,

$$Al + 3HCl \rightarrow AlCl_3 + 1\cdot5H_2$$

Aluminium chloride is a white solid which volatilizes on heating. In a sealed tube, the solid melts under pressure at 193–194°C. When rapidly heated, a sample of aluminium chloride melts and then, at 183°C, boils.

The anhydrous compound is very hygroscopic and reacts with water, the hydrolysis being accompanied by the evolution of much heat.

From vapour density measurements at elevated temperatures, it has been deduced that Al_2Cl_6 molecules are present in the vapour below 44°C, that dissociation according to $Al_2Cl_6 \rightleftharpoons 2AlCl_3$ occurs between 440° and 600°C and that decomposition occurs beyond 1100°C.

Aluminium chloride hexahydrate is made by dissolving aluminium in concentrated hydrochloric acid followed by evaporation of the solution or saturation of it with hydrogen chloride gas whereupon crystals of $AlCl_3 \cdot 6H_2O$ are formed. These cannot be converted to the anhydrous compound by thermal dehydration because this results in the loss of hydrogen chloride as well as water and the formation of a basic aluminium chloride.

Reactions

Aluminium chloride forms addition compounds with very many other substances.

With ammonia, a range of solid ammines of general formula $AlCl_3 \cdot nNH_3$ is known. When excess dry ammonia is passed over sublimed aluminium chloride, the hexammine ($n = 6$) is produced. This compound loses ammonia on heating and various lower ammines ($n = 5, 3$ or 2) are produced. The reaction between equimolar amounts of aluminium chloride and ammonia gives the monoammine, $AlCl_3 \cdot NH_3$ (m.p. = 130°C).

When aluminium chloride is heated with other metal chlorides or ammonium chloride, a tetrachloroaluminate is produced. Thus, aluminium and sodium chlorides react between 200° and 240°C under nitrogen to give $NaAlCl_4$ (m.p. = 156°C).

At least three compounds between $AlCl_3$ and $POCl_3$ are known to exist. They have been identified as $AlCl_3POCl_3$, $AlCl_32POCl_3$ and $AlCl_36POCl_3$ from the melting-point/composition diagram for mixtures of the two components. All are white solids which are decomposed by water. Coordination occurs through the oxygen, $Cl_3PO \rightarrow AlCl_3$, although possibly the ionic form $Cl_2PO^+AlCl_4^-$ exists in solution.

Addition compounds with other inorganic molecules such as H_2S, SO_2, $SOCl_2$, $SeCl_4$, $TeCl_4$, ICl and ICl_3 have been identified and characterized by physico-chemical studies.

A complex of stoichiometry $AlCl_3PCl_5$ is obtained by reaction with PCl_5. This is a white solid, m.p. 380°C, which dissolves in nitrobenzene to give a conducting solution which probably contains $AlCl_4^-$ and PCl_4^+ ions. The $AlCl_4^-$ ion results from the coordination of chloride ion through one of its lone pair of electrons by the $AlCl_3$ molecule: $AlCl_3 + Cl^- \rightarrow AlCl_4^-$. This exemplifies the behaviour of $AlCl_3$ as a Lewis acid, a feature which is of particular significance in its reactions with organic compounds.

One of the most important uses of aluminium chloride is as a catalyst for Friedel–Crafts reactions. For example, it catalyses the alkylation of aromatic compounds using alkyl halides, alcohols or olefins:

$$Ar-H + RX \xrightarrow{AlCl_3} Ar-R + HX$$
$$Ar-H + ROH \longrightarrow Ar-R + H_2O$$
$$Ar-H + R-CH=CH_2 \longrightarrow \underset{\underset{Ar}{|}}{R-CH-CH_3}$$

Similarly, it is effective in the acylation of aromatic compounds using an acid halide, acid anhydride, ester or acid:

$$Ar-H + RCOX \xrightarrow{AlCl_3} Ar-R + HX$$
$$Ar-H + (RCO)_2O \longrightarrow ArCOR + RCOOH$$

Other metal halides show similar catalytic properties. These include $AlBr_3$, $BeCl_2$, $CdCl_2$, $ZnCl_2$, BF_3, BCl_3, BBr_3, $GaCl_3$, $TiCl_4$, $ZrCl_4$, $SnCl_4$, $SbCl_3$, $SbCl_5$ and $FeCl_3$. These, like $AlCl_3$, are Lewis acids, the metal atom in each case accepting a pair of electrons from a Lewis base.

Aluminium chloride is the most commonly used catalyst in Friedel–Crafts reactions. The catalytic action is associated with the formation of a complex with donor molecules such as alkyl or acyl halides.

In the case of alkyl halides, the complex is formed thus:

$$RX + AlCl_3 \rightarrow RXAlCl_3$$

It has been suggested that such complex formation weakens the C–X bond in the alkyl halide so that ionization to reactive carbonium ions is promoted, e.g.

$$RCl + AlCl_3 \rightleftharpoons R^+[AlCl_4]^- \rightleftharpoons R^+ + [AlCl_4]^-$$

$$ArH + R^+ \rightarrow Ar-R + H^+$$

$$H^+ + [AlCl_4]^- \rightleftharpoons HCl + AlCl_3$$

The carbonium ions then effect the electrophilic substitution in the aromatic nucleus.

Aluminium chloride readily dissolves in liquid organic compounds which contain donor atoms or groups, e.g. alcohols, ethers, ketones and acids. Freezing-point depression measurements show that in these solutions, the aluminium chloride is monomeric. This form is stabilized by the formation of a 1:1 addition complex with the solvent molecule, the aluminium atom again, as in the above examples, accepting two electrons from the donor atom to complete its valence shell of eight electrons.

Some complexes between acyl halides, RCOX, and aluminium chloride have been isolated and characterized. Their infrared spectra have been interpreted as showing that both the coordination complex, $RCXO \rightarrow AlCl_3$, and the ionic form, $RCO^+AlCl_4^-$, of the same compound can co-exist. In the case of the addition compound with benzoyl chloride, it appears from chlorine exchange studies, that the ions are present in trace amounts only. The complex, $CH_3COClAlCl_3$, appears to be a mixture of $CH_3CClO \rightarrow AlCl_3$ and $CH_3CO^+AlCl_4^-$. The oxocarbonium ion, CH_3CO^+, is the effective acylating agent in the Friedel–Crafts reaction.

Structure

The Al_2Cl_6 molecule in the vapour state is known, from electron diffraction studies,[1,2] to have the stereochemistry of two tetrahedra sharing an edge. The two aluminium atoms are located centrally in these tetrahedra (Fig. 3.1). Interatomic distances (nm) reported by Palmer and Elliott are:

$$\text{Average of four } AlCl_t \text{ distances} = 0.206 \pm 0.004$$

$$\text{Average of four } AlCl_b \text{ distances} = 0.221 \pm 0.004$$

$$Cl_b, Cl_b = 0.283 \pm 0.01; \; Al, Al = 0.341 \pm 0.02$$

The distance between the two bridging chlorines is short enough to result in a flattening of the two tetrahedra along the Cl_b–Cl_b direction. Akishin and co-workers' values for the $AlCl_t$ and $AlCl_b$ distances are 0.204 ± 0.002 and 0.224 ± 0.002 nm respectively.

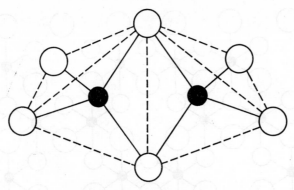

Fig. 3.1. The structure of the Al_2Cl_6 molecule.

It is interesting to note that the electron diffraction pattern of aluminium bromide vapour is very similar to that of aluminium chloride suggesting both compounds have the same structure.

The vapour of aluminium chloride at high temperatures contains $AlCl_3$ molecules. These are trigonal planar molecules of D_{3h} symmetry[3] and their formation by dissociation of the dimer necessitates a change in stereochemistry of the aluminium. In $AlCl_3$, the Al,Cl bond length $= 0.206 \pm 0.001$ and the Cl,Cl distance $= 0.353 \pm 0.001$ nm.

An X-ray diffraction study of molten aluminium chloride shows that Al_2Cl_6 molecules are also present in this phase.[4] Formation of crystalline aluminium chloride from the melt or vapour leads to a breakdown of the dimeric molecules and the formation of a layer lattice composed of close-packed chlorines with aluminium atoms occupying two-thirds of the octahedral holes.[5] Thus solid aluminium chloride forms a typically ionic lattice − the layers are of exactly the same type as found in chromium(III) chloride (Fig. 3.2). The whole structure is slightly distorted and crystallizes in monoclinic crystals. The unit cell has dimensions:

$$a = 0.592; \quad b = 1.022; \quad \text{and} \quad c = 0.616 \text{ nm}; \quad \beta = 108°$$

The electrical conductivity of solid aluminium chloride increases[2] with temperature, a particularly sharp increase occurring at about 140°C. This is the same temperature at which an endothermic change is observed in the solid. There appears to be no significant structural change accompanying these effects so it has been proposed that increasing thermal vibrations in the lattice cause a reduction in the van der Waals interaction between the layers, a regrouping of these and a weakening of the Al–Cl bonds within them. As a result, some of the aluminium ions become distributed randomly in tetrahedral sites and the mobility of such ions increases the electrical conductivity. The conductivity decreases sharply at the melting-point and is almost zero for the liquid. On fusion, the bonds between the layers in aluminium chloride are further weakened and unrestricted movement of aluminium ions into tetrahedral sites can occur to form Al_2Cl_6 molecules. The melt composed of these is, of course, nonconducting.

From a structural point of view, aluminium chloride occupies an intermediate position between the fluoride (m.p. = 1290°C) which has an ionic lattice, and aluminium bromide (m.p. = 97.5°C) which exists as dimeric molecules in the solid state as well as in the liquid and vapour. The changes observed in the Raman spectrum of aluminium chloride on solidification confirm that structural changes accompany this.

29

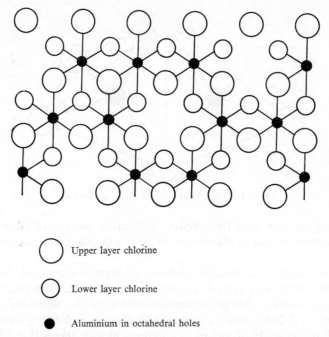

○ Upper layer chlorine

◔ Lower layer chlorine

● Aluminium in octahedral holes

FIG. 3.2. Layer lattice structure of aluminium chloride.

Vibrational Spectrum

Dimeric aluminium chloride, having a similar structure to diborane, has the same symmetry, D_{2h}, and the same number of fundamental vibrations (Fig. 1.2).

The observed frequencies in the Raman spectrum of the liquid and their assignments are given in Table I. The original measurements of Gerding and Smit[6] have been confirmed more recently by Pershina and Raskin.[7] Seven out of the nine permitted Raman fundamentals have been observed. Calculated values[8] for those not observed are $\nu_7 = 200$ and $\nu_{15} = 152$ cm^{-1}.

The infrared spectrum of Al_2Cl_6 vapour in the region 1200–325 cm^{-1} shows[9] three bands, at 625, 420 and 484 cm^{-1} (assigned respectively as ν_8, ν_{13} and ν_{16}). The remaining five infrared active bands, ν_9, ν_{10}, ν_{11}, ν_{17} and ν_{18}, are all expected to lie below 325 cm^{-1}.

In a theoretical treatment of the vibrational spectrum of Al_2Cl_6 using a modified Urey–Bradley force field, Onishi and Shimanouchi[8] have calculated values of the force constants. The bridge stretching force constant, $K_b(Al–Cl_b) = 105$ Nm^{-1}, and the terminal constant, $K_t(Al–Cl_t) = 235$ Nm^{-1}. The terminal chlorines can be regarded as forming normal Al–Cl single bonds and these figures then may be interpreted as showing that, as in diborane, the bridge bonds are much weaker than the terminal bonds.

One of the infrared active vibrations for the monomeric molecule has been observed[9] in the vapour of aluminium chloride at 900°C. The band, which occurs at 610 cm^{-1}, has been assigned as the stretching mode. Two other infrared active bands are expected for a planar $AlCl_3$ molecule (symmetry D_{3h}). These, primarily associated with bending modes, should occur in the far infrared region well below 610 cm^{-1}.

TABLE I

The Raman Spectrum of Dimeric Aluminium Chloride (frequencies in cm^{-1})

Vibration	Observed		Calculated
	Gerding and Smit[6]	Pershina and Raskin[7]	Onishi and Shimanouchi[8]
ν_1	506	510	513
ν_2	340	339	339
ν_3	217	219	222
ν_4	112	119	106
ν_6	284	282	287
ν_7	—	—	200
ν_{11}	606	605	598
ν_{12}	163	164	157
ν_{15}	—	—	152

Aqueous Solutions of Aluminium Chloride

The physico-chemical properties of aqueous solutions of many metal salts show clearly that the metal ions are hydrated, with a definite number (the hydration number of the ion) of water molecules bound directly to each ion to form a primary coordination shell. In the case of a highly charged ion like Al^{3+}, the interaction between this and coordinated water should be particularly strong. Solutions of aluminium compounds like the chloride have therefore been particularly attractive systems to study.

N.M.R. spectroscopy has proved to be a valuable experimental technique in this field, for one can exploit the resonance due to ^{17}O ($I = 5/2$) or ^1H ($I = 1/2$). It is expected that the N.M.R. absorption for either of these nuclei should occur at substantially different fields depending on whether it is in a combined or a solvent water molecule.

In the case of ^{17}O resonance, it has been observed[10] that the addition of paramagnetic ions to an aqueous solution of a complex ion containing coordinated water alters the field strength at which ^{17}O in the free water molecules resonates, but not that in the combined molecules. For example, the presence of Co^{2+} in an aqueous solution containing $[(NH_3)_5Co^{III}(H_2O)]^{3+}$ shifts to a lower field the position of the resonance peak for ^{17}O in the solvent water molecules leaving in the complex unchanged. The shift results from the formation of hydrated Co^{2+} ions and rapid exchange of molecules between these and the bulk of the solvent. Very few molecules of water in the complex ion are likely to be bonded to Co^{2+} and so there is virtually no contribution made by Co^{2+} to the average field experienced by these.

The natural abundance of ^{17}O is 0·04%. This is too low to permit a precise measurement of the hydration number of Al^{3+} using the above method for distinguishing combined and uncombined water. However, using water in which the ^{17}O content had been enriched to 11·48%, Connick and Fiat[11] found the hydration number of Al^{3+} to be 5·93 (an average of four separate measurements) indicating that the species predominating in aqueous solution is $Al(H_2O)_6^{3+}$.

The proton magnetic resonance spectrum of aqueous aluminium chloride at room temperature shows only one peak, an average of proton resonances as a result of rapid exchange, which makes it impossible to distinguish between coordinated and solvent water. However, the direct observation of separate proton resonance peaks for bulk water and water in the coordination shell of Al^{3+} has been reported[12] on solutions of aluminium chloride at low temperatures. The freezing-point decreases sharply with concentration and a solution of 133 g of $AlCl_3$ in 324 g of water is still liquid at $-47°C$. From the integrated areas of the resonance peaks, the coordination number of aluminium in this solution has been calculated as 6·0.

Solutions of aluminium chloride are acidic because the complex ion, $Al(H_2O)_6{}^{3+}$, tends to lose a proton according to:

$$Al(H_2O)_6{}^{3+} \rightleftharpoons Al(H_2O)_5(OH)^{2+} + H^+$$

The equilibrium constant for this dissociation is $K = 1·12 \times 10^{-5}$.

Energetics of Adduct Formation by Aluminium Chloride

Examples of adduct formation by aluminium chloride acting as a Lewis acid have already been given. The overall energy change associated with such reactions is determined by a number of factors. If we consider that aluminium chloride reacts with a Lewis base B, both reactants being initially in condensed states, to form the adduct $BAlCl_3$, the reaction can be subdivided into several steps, each of which has a particular energy change:

(i) Vaporization of aluminium chloride and B.

(ii) Breakdown of Al_2Cl_6 into $AlCl_3$ molecules. This involves the breakage of Al–Cl bonds and a change in the coordination of aluminium from tetrahedral to trigonal planar.

(iii) Reorganization of the bonding orbitals in aluminium to give the pyramidal configuration which the $AlCl_3$ group has in the adduct. In the trigonal planar $AlCl_3$ molecule, the aluminium atom forms three σ-bonds and has an unfilled p_z orbital perpendicular to the plane which can accept electrons from a filled p_z orbital on chlorine thus forming a π-bond. When orbital reorganization occurs, such π-bonding is reduced in strength or may disappear altogether. There will also be a change in the strength of the Al–Cl σ-bonds resulting from the different hybridization of aluminium. A certain amount of reorganization energy is required to effect the change to pyramidal configuration and this is believed to contribute significantly to the overall energy change accompanying adduct formation.

(iv) Combination of 'reorganized' $AlCl_3$ with B to form the adduct in the gas phase.

(v) Condensation of the adduct to the liquid or solid state.

Experimental data are available[13] on reactions of the type: $B(g) + AlX_3(g) \rightarrow BAlX_3(c)$ where X = Cl, Br or I, and B is a base like pyridine, ammonia, trimethylamine or triethylamine. For $AlCl_3$ the enthalpies of reaction (in $kJ\ mol^{-1}$) are in the range 250–285, for $AlBr_3$ in the range 255–265 and for AlI_3, 420–450. The relevant heats of sublimation of complexes between a given amine and different halides are little different from one another. These figures could therefore be interpreted as showing that, for the halides in their reorganized pyramidal configurations, the chloride and bromide have comparable acceptor power but that of the iodide is much greater. It is very unlikely that AlI_3 would be the best acceptor of these three and it appears that an alternative explanation, in terms of the reorganization energy, should be sought.

The magnitude of this energy should be greatest where strong π-bonds in the planar molecule are broken on rehybridization to the pyramidal configuration in the adduct. The metal–halogen π-bonding is not necessarily completely destroyed on conversion to a pyramidal shape. Indeed, the distinction between σ- and π-molecular orbitals disappears[15] with the loss of the mirror plane of symmetry inherent in planar trigonal molecules. However, it can be assumed that, on adduct formation, the p_z orbital on aluminium becomes fully involved in σ-bonding with the ligand and that the Al–X π-bonding no longer occurs.

Estimated values for the reorganization energy are given in Table II. The two sets of values do not agree, but, significantly, the trend in both cases is towards the lowest energy for the iodide. This is as expected, for π-bonding should be greatest in the chloride and least in the iodide.

TABLE II

Estimated Values for the Reorganization Energy of Aluminium Halides

Reference	Reorganization energy (kJ mol^{-1})		
	$AlCl_3$	$AlBr_3$	AlI_3
Eley and Watts[13]	360	300	175
Cotton and Peto[14]	132	116	78

The above arguments based on the importance of reorganization energy have also provided a logical explanation of the fact that, of the three boron trihalides (BF_3, BCl_3 and BBr_3), BF_3 appears to be the weakest acceptor molecule. At first sight, we might expect that the high electronegativity of fluorine would promote electron withdrawal from the boron and lead to marked acceptor properties. In fact, the B–F π-bonding in BF_3 is extensive, the reorganization energy is consequently large and the overall enthalpy of formation of the BF_3 adduct with a given base is lower than that for the other halides.

Bonding

Some conclusions on the nature of the bonding in solid aluminium chloride have been reached from observations of the quadrupole splitting of the ^{27}Al nucleus. For this, $I = 5/2$, and its nuclear magnetic resonance shows quadrupole splitting. The nuclear quadrupole coupling constant of ^{27}Al in anhydrous solid aluminium chloride has been found to be 472 ± 24 kHz. On the assumption of a completely ionic structure, the electric field gradient at the aluminium nucleus has been calculated[16] by the direct summation of effects due to the surrounding ions (regarded as point charges). This leads to an estimated value of $-2 \cdot 78$ MHz for the quadrupole coupling constant of ^{27}Al. The discrepancy between the experimental and estimated values has been attributed to the neglect of covalent bonding and the relatively large effect on the electric field gradient which a small amount of this is believed to produce. For a perfectly octahedral coordination of aluminium, the coupling constant due to covalent bonding with six nearest neighbours would be zero. In fact, there is some distortion of the octahedron and the effect of this on the coupling constant has to be considered. The nuclear quadrupole coupling constant calculated for completely covalent bonding is $+4 \cdot 05$ MHz. As the experimentally determined constant is between these two limiting theoretical values, it is possible to estimate the extent of covalent bonding in the solid. This is approximately $1/3$, the remaining $2/3$ being ionic. The charge carried by each aluminium ion is therefore $+2$.

When aluminium chloride melts, the weak bonding between the layers in the crystal lattice breaks down first and, within the layers, the aluminium ions migrate from octahedral to tetrahedral holes. This leads to the formation of Al_2Cl_6 molecules. The process is reversible and there must be only a small energy difference between the two structures to account for the effect of temperature on their relative stabilities.

Kato, Yamaguchi and Yonezawa[17] have treated the problem of the electronic structure of aluminium chloride dimer by the extended Hückel method proposed by Hoffmann.[18] This represents the molecular orbitals in terms of linear combinations of the atomic orbitals of all the valence electrons and does not presuppose particular hybridizations or orientations of these orbitals. They concluded that the bonds between aluminium and the terminal chlorines are strong and little affected by dissociation of the dimer, that the bonds between aluminium and the bridge chlorines are much weaker and that there is only weak direct metal–metal bonding in the molecule. The dimer is regarded as being electron deficient in the same way as diborane.

References

1. PALMER, K. J., and ELLIOTT, N., *J. Amer. Chem. Soc.* **60**, 1852 (1938).
2. AKISHIN, P. A., RAMBIDI, N. G., and ZASORIN, E. Z., *Kristallografiya* **4**, 186 (1959).
3. ZASORIN, E., and RAMBIDI, N. G., *Zhur. Strukt. Khim.* **8**, 391 (1967).
4. HARRIS, R. L., WOOD, R. E., and RITTER, H. L., *J. Amer. Chem. Soc.* **73**, 3151 (1951).
5. KETELAAR, J. A. A., MACGILLAVRY, C. H., and RENES, P. A., *Rec. Trav. Chim.* **66**, 501 (1947).
6. GERDING, H., and SMIT, E., *Z. Phys. Chem.* B **50**, 171 (1941).
7. PERSHINA, E. V., and RASKIN, S. S., *Optika i Spekt.* **13**, 488 (1962).
8. ONISHI, T., and SHIMANOUCHI, T., *Spectrochim. Acta* **20**, 325 (1964).
9. KLEMPERER, W., *J. Chem. Phys.* **24**, 353 (1956).
10. JACKSON, J. A., LEMONS, J. F., and TAUBE, H., *J. Chem. Phys.* **32**, 553 (1960).
11. CONNICK, R. E., and FIAT, D. N., *J. Chem. Phys.* **39**, 1349 (1963).
12. SCHUSTER, R. E., and FRATIELLO, A., *J. Chem. Phys.* **47**, 1554 (1967).
13. ELEY, D. D., and WATTS, H., *J. Chem. Soc.* 1319 (1954).
14. COTTON, F. A., and PETO, J. R., *J. Chem. Phys.* **30**, 993 (1959).
15. ARMSTRONG, D. R., and PERKINS, P. G., *J. Chem. Soc.* A 1218 (1967).
16. CASABELLA, P. A., and MILLER, N. C., *J. Chem. Phys.* **40**, 1363 (1964).
17. KATO, H., YAMAGUCHI, K., and YONEZAWA, T., *Bull. Chem. Soc. Jap.* **39**, 1377 (1966).
18. HOFFMANN, R., *J. Chem. Phys.* **39**, 1397 (1963).

CHAPTER 4

NITROGEN DIOXIDE

Introduction

The oxides of nitrogen constitute a unique series for there is no other group of binary compounds which shows such a diversity in number, structures and physico-chemical properties.

Nitrogen dioxide, NO_2, is the most well known of these and, for a number of reasons, it is a compound of special interest. It is made in vast quantities annually for the manufacture of nitric acid. In the liquid state it is used as a solvent medium for the preparation of compounds which cannot be made in more conventional solvents. Nitrogen dioxide in the gas phase shows a remarkably complex electronic spectrum and the theoretical interpretation of this is by no means settled. The molecule contains an odd number of electrons and is one of the simplest of free radicals. It has therefore been an attractive subject for the investigation of the effects of electron spin in molecules.

NO_2 forms a diamagnetic dimer, dinitrogen tetroxide, N_2O_4. The chemical equilibrium $2NO_2 \rightleftharpoons N_2O_4$ has been extensively investigated. Several dimers appear to be capable of existence depending on whether the NO_2 molecules are linked via nitrogen or oxygen. The dimer which has been most fully characterized contains an unusually long N,N bond and various theoretical explanations have been proposed to account for this feature.

Preparation

Nitrogen dioxide has been known for many centuries. Its formation would have been observed in the first preparations of nitric acid from nitre. An early description[1] by Robert Boyle appears in his work, *Experiments and Considerations Touching Colours*: 'we observe, in the distilling of pure salt-peter, that, at a certain season of the operation, the body, although it seem either crystalline or white, affords very red fumes' and 'though ... by skill and care a reddish liquor may be obtained from nitre, yet the common spirit of it, in the making even of which store of these red fumes are wont to pass over into the receiver, appears not to be at all red'.

Many of the properties of nitrogen dioxide were first examined by Joseph Priestley in the course of his unique studies on gases, the results of which were published between 1774 and 1790. He prepared the gas by the action of heat on lead nitrate. For many years thereafter this was used as the standard preparative method and, indeed, it is still a convenient source of the gas in the laboratory. One problem in working with nitrogen dioxide is its reactivity towards liquids like water and mercury. These cannot therefore be used in manipulations involving the gas. To overcome this difficulty, Priestley developed the technique of displacement of air from a glass vessel as a means of collecting the gas. He was able to isolate nitrogen dioxide and study in detail its reactions with water, essential oils, caustic alkalis and sulphuric acid.

Nitrogen dioxide can be readily prepared by the action of concentrated nitric acid (65% and over) on copper metal at room temperature. The gaseous product also contains nitric and nitrous oxides.

The compound may be made in a state of high purity[2] by heating lead nitrate, previously dried at 120°C for several hours, to 450°C.

$$2Pb(NO_3)_2 \rightarrow 4NO_2 + O_2 + 2PbO$$

The gas is passed through phosphorus pentoxide and condensed in a receiver cooled by an acetone/solid carbon dioxide mixture. Oxygen is pumped out of the system and any of the gas which may be trapped in the solid nitrogen dioxide is removed by the sublimation of this under vacuum.

Nitrogen dioxide is made on an industrial scale by the reaction between nitric oxide and oxygen.

$$2NO + O_2 \rightarrow 2NO_2$$

Nitric oxide itself can be prepared directly from the nitrogen and oxygen of the air or indirectly by catalytic synthesis and oxidation of ammonia.

Properties

The empirical formula of nitrogen dioxide was established independently in 1816 by Dulong and Gay-Lussac. In the gas phase, the species chiefly present is the monomer. In the liquid and solid phases it is almost entirely N_2O_4.

The degree of dissociation of N_2O_4 has been determined by pressure and volume measurements on a known quantity of gas at various temperatures. At the boiling-point (21·15°C) the vapour is brown and contains about 16% by weight of NO_2. With increase in temperature, the colour of the vapour intensifies and it becomes black around 150°C due to further dissociation. Above 160°C, decomposition to nitric oxide and oxygen becomes appreciable.

In the liquid state, N_2O_4 is yellow at the triple point ($-11\cdot2°C$) and its colour darkens as the temperature is raised. Since NO_2 absorbs radiation over a wide range of the visible spectrum whereas N_2O_4 is transparent over almost all this wavelength range, the colour of the liquid serves as an indication of the concentration of monomer.

Spectrophotometry of the liquid has shown[3] it is 0·12% dissociated at 21·15°C at atmospheric pressure and only 0·038% at $-11\cdot2°C$.

The most recent spectrophotometric study of the NO_2/N_2O_4 equilibrium in the gas phase is that carried out by Harris and Churney.[4] They evaluated the dissociation constant, K, by measurements on gaseous mixtures of the absorbance by NO_2 of mercury vapour radiation of wavelength 546·1 nm. These, coupled with calculated values of the N_2O_4 content from the known vapour pressure of N_2O_4 solid, gave the values for K shown in

Table I. This also includes, for comparison, some of Hisatsune's values,[5] calculated from molecular dimensions and spectroscopic data. The enthalpy change, ΔH^0_{25}, for the dissociation reaction in the gas phase at 25 °C was estimated as 57·06 kJ mol^{-1} by Hisatsune.

The dimerization of NO_2 in solution has been examined by N.M.R. spectroscopy.[6] The paramagnetism of NO_2 makes it possible to apply this technique to the measurement of the volume magnetic susceptibility of solutions of NO_2 and N_2O_4 at equilibrium in solvents like cyclohexane, carbon tetrachloride and acetonitrile. From the known magnetic

TABLE I

Equilibrium Constants for NO_2/N_2O_4 Equilibrium in the Gas Phase

Experimental (Harris and Churney[4]):			
K (kNm^{-2})	16·46	22·22	32·0
Temperature (°C)	26·55	30·54	35·73
Calculated (Hisatsune[5]):			
K	15·30	17·63	
Temperature	25	26·84	

TABLE II

Equilibrium Constants for NO_2/N_2O_4 in Solution

Solvent	K_{298} (mol kg^{-1})	ΔH^0_{25} (kJ mol^{-1})
Cyclohexane	(1·77 ± 0·1) 10^{-4}/density	61·1 ± 2·1
Carbon tetrachloride	(1·78 ± 0·1) 10^{-4}/density	61·1 ± 4·2
Acetonitrile	(0·3 ± 0·1) 10^{-4}/density	66·9 ± 6·3

susceptibilities of pure NO_2 and N_2O_4, the values of K at 25 °C in the three solvents were found (Table II). The extent of dissociation of N_2O_4 in a non-coordinating solvent is clearly much less than in the gas phase. Since the enthalpy of dissociation increases only slightly in going from the vapour phase to the solvent environment, the reduction in dissociation of N_2O_4 in solution must be primarily due to a much lower entropy change in this phase.

Dinitrogen Tetroxide Solvent System

Before the present century, almost all preparative work in inorganic chemistry which involved reactions in solution, used water as the solvent medium. Recent studies with a number of inorganic non-aqueous solvents have made clear the limitations of water as a solvent. Many compounds cannot be prepared from aqueous solution although they are perfectly stable when made under anhydrous conditions. Liquid dinitrogen tetroxide is a useful solvent for the preparation of many nitrogen-containing compounds which are difficult or impossible to make in aqueous solution.

The dielectric constant, ε, of liquid N_2O_4 is 2·42 at 15°C (cf. benzene, for which $\varepsilon = 2·60$) and its specific conductivity is low. It may therefore be concluded that very few ions are present in the pure liquid.

Dinitrogen tetroxide undergoes self-ionization according to:

$$N_2O_4 \rightleftharpoons NO^+ + NO_3^-$$

37

The concentrations of nitrosonium (NO^+) and nitrate (NO_3^-) ions in the pure liquid are extremely low and their presence is very difficult to detect by conventional methods. However, an elegant demonstration that some nitrate ions must be present has come from exchange studies using the ^{15}N isotope. Complete exchange of ^{15}N is observed[7] in a solution of tetramethylammonium nitrate $[(CH_3)_4N]^+[^{15}NO_3]^-$ in liquid N_2O_4.

Self-ionization of N_2O_4 is enhanced by dissolving it in solvents of high dielectric constant. For example, the conductivity of N_2O_4 dissolved in nitromethane ($\varepsilon = 37$) is greater than that of pure N_2O_4 due to increased ionization. In acetic acid/nitric acid mixtures, N_2O_4 behaves as a weak electrolyte and in pure nitric acid its conductivity is that expected for a moderately strong electrolyte. In pure sulphuric acid ($\varepsilon \sim 110$), ionization to NO^+ is complete. The nature of the other species present in this acid depends on the concentration of N_2O_4. For up to 2·6 mole % N_2O_4 all the N_2O_4 is ionized according to:

$$N_2O_4 + 3H_2SO_4 \rightleftharpoons NO^+ + NO_2^+ + H_3O^+ + 3HSO_4^-$$

At higher concentrations, molecular nitric acid is formed.

$$N_2O_4 + H_2SO_4 \rightleftharpoons NO^+ + HSO_4^- + HNO_3$$

Above 16 mole % N_2O_4, nitrosonium bisulphate, $[NO]^+[HSO_4]^-$, crystallizes from solution.

Aromatic compounds are nitrated by solutions of N_2O_4 in sulphuric acid, the effective nitrating agent being NO_2^+. The existence of this ion and NO^+ in sulphuric acid solutions has been demonstrated by observation of their characteristic Raman spectra.[8]

The formation of nitrosonium salts is a feature of the chemistry of N_2O_4. When N_2O_4 is dissolved in a strong acid, HX, HNO_3 is formed leaving X^- and NO^+ ions. The protonation of NO_3^- effectively reduces its concentration and further ionization of N_2O_4 proceeds. If the solubility product of the salt, NO^+X^-, is exceeded, this is precipitated. In addition to the bisulphate, other nitrosonium salts which can be prepared in this way are $NOClO_4$ and $NOReO_4$.

The tetrafluoroborate, $NOBF_4$, is formed when BF_3 is passed into a solution of N_2O_4 in pure nitric acid. Different products are obtained when BF_3 reacts with pure N_2O_4. These have the stoichiometry $N_2O_4 \cdot BF_3$ and $N_2O_4 \cdot 2BF_3$ and are formulated as ionic compounds, $[NO_2]^+[BF_3NO_2]^-$ and $[NO_2]^+[N(OBF_3)_2]^-$ respectively. These reactions suggest that ionization of N_2O_4 has occurred according to:

$$N_2O_4 \rightleftharpoons NO_2^+ + NO_2^-$$

This reaction is the first stage of the self-ionization of N_2O_4. It proceeds no further in the presence of a strong electron acceptor like BF_3, the NO_2^- ions donating electrons to form $[BF_3NO_2]^-$ and $[N(OBF_3)_2]^-$. In the absence of strong electron acceptors, the second stage of self-ionization occurs, in which an oxygen atom is transferred from NO_2^+ to NO_2^- to give NO^+ and NO_3^-, the more commonly observed products of ionization.

Few inorganic salts are soluble in dinitrogen tetroxide and in some ways it resembles organic solvents like benzene in its properties. Non-metallic elements like bromine and iodine dissolve freely and N_2O_4 is miscible with covalent inorganic compounds like CS_2 and $SiCl_4$. Many organic compounds dissolve readily in N_2O_4 without reaction and may be recrystallized using it as solvent. N_2O_4 forms addition compounds with tertiary amines, heterocyclic nitrogen compounds, ethers, nitriles and ketones. These are of general formulae $N_2O_4 \cdot B$ and $N_2O_4 \cdot 2B$, where B is the organic base. 1:2 complexes have been found for all bases studied and in some cases, 1:1 complexes are also known.

Dinitrogen tetroxide is a useful solvent medium for the preparation of metal nitrates and nitrato-complexes. For example, sodium, potassium, silver, zinc, mercury and lead react with N_2O_4 to give the corresponding metal nitrate and nitric oxide, e.g.

$$K + N_2O_4 \rightarrow KNO_3 + NO$$

With a weakly electropositive metal like zinc, reaction continues further to form an insoluble nitrosonium salt $[NO]_2^+[Zn(NO_3)_4]^{2-}$. The reaction between metals and N_2O_4 is facilitated by dilution of the latter with organic solvents, such as nitromethane and ethyl acetate. In the case of nitromethane, self-ionization of N_2O_4 is promoted by the high dielectric constant of the organic solvent. The dielectric constant of ethyl acetate is too low ($\varepsilon = 6\cdot02$ at 25°C) for this to be the effective mechanism, and in this case it is likely that self-ionization is promoted by the solvation and stabilization of NO^+ with ethyl acetate molecules. The use of mixed solvents like these is a valuable practical technique for the preparation of anhydrous metal nitrates such as copper(II) nitrate (p. 220).

Nitrato-complexes are formed when some metal nitrates are dissolved in N_2O_4. For example, zinc nitrate gives a compound of stoichiometry $Zn(NO_3)_2 \cdot 2N_2O_4$, and uranyl nitrate gives $UO_2(NO_3)_2 \cdot N_2O_4$. In fact, these compounds are nitrosonium salts and are formulated as $[NO]_2^{2+}[Zn(NO_3)_4]^{2-}$ and $[NO]^+[UO_2(NO_3)_3]^-$ respectively. The complex anion $[Zn(NO_3)_4]^{2-}$ is the analogue of the ions $[Zn(OH)_4]^{2-}$ and $[Zn(NH_2)_4]^{2-}$ which are found in the water and liquid ammonia solvent systems respectively.

Reactions of Nitrogen Dioxide with Water and Alcohols

The reaction between nitrogen dioxide gas and water has been extensively studied because of its importance in the manufacture of nitric acid. Nitrous and nitric acids are formed in equimolar quantities in the presence of an excess of water:

$$2NO_2 + H_2O \rightarrow HONO + HONO_2$$

The nitrous acid formed slowly decomposes at low temperatures into nitric oxide and nitric acid:

$$3HONO \rightarrow H_2O + HONO_2 + 2NO$$

The nitric oxide formed reduces the equilibrium concentration of nitric acid in solution which can be attained with a given pressure of nitrogen dioxide. Study of the reaction kinetics has shown that the rate of formation of nitric acid much stronger than 50% is too slow at atmospheric pressure to be worthwhile in practice. Higher concentrations of nitric acid can be made by using a mixture of air and nitrogen dioxide for reaction with water. Then the nitric oxide formed reacts with oxygen to give more nitrogen dioxide.

Similar reactions occur with alcohols, ROH:

$$ROH + 2NO_2 \rightleftharpoons RONO + HONO_2$$

In solution, two liquid layers are produced, one being almost pure nitrite ester, the other concentrated nitric acid. It has been established, using alcohols labelled with ^{18}O, that the C,O bond is not broken during the reaction of NO_2 with an alcohol. Reaction thus proceeds via the formation of NO^+ and NO_3^-.

Spectroscopic Studies and Structure of NO_2

The shape and dimensions of the NO_2 molecule have been determined by infrared spectroscopy,[9] microwave spectroscopy[10] and electron diffraction.[11] The results are summarized in Table III.

TABLE III

Molecular Parameters of Nitrogen Dioxide as determined Spectroscopically and by Electron Diffraction

	Moore[9]	Bird[10]	Claesson et al.[11]
N,O bond length (nm)	0·1188 ± 0·0004	0·1197	0·120 ± 0·0002
O–N–O (deg)	134·07 ± 0·25	134·25	132 ± 3

Force constants (Nm^{-1})	Arakawa and Nielsen[15]	Bird et al.[19]
f_d	1092·7 ± 6·5	1104 ± 5
f_α	112·5 ± 0·3	110·9 ± 1
f_{dd}	203·8 ± 0·5	214 ± 5
$f_{d\alpha}$	39 ± 20	48·1 ± 1·4

They show that NO_2 is a bent molecule of C_{2v} symmetry. The infrared spectrum of the gas has been interpreted on the basis of this model and the rotational fine structure of the vibrational bands is typical of a symmetric top molecule.

The symmetry properties of the fundamental vibrations of the NO_2 molecule are: v_1, symmetric stretching; v_2, symmetric bending; and v_3, antisymmetric stretching vibrations. All three are Raman and infrared active. Moore[9] has made the following assignments from his study of the infrared spectrum of NO_2: $v_1 = 1322·5 ± 0·2$ cm^{-1}; $v_2 = 750·9 ± 0·4$ cm^{-1}; and $v_3 = 1616 ± 0·2$ cm^{-1}. The frequencies for v_1 and v_2 are in good agreement with the values 1321·1 and 751·1 cm^{-1} respectively which have been deduced from the ultraviolet absorption spectrum of NO_2 gas.[12] The absorption band at 249·1 nm shows well-developed branches corresponding with rotational fine structure. Weaker bands of very similar structure are centred at 1321·1 and 751·1 cm^{-1} on the long wavelength side of this band. These must arise from different vibrational states of the molecule and it is reasonable to identify their separations from the main absorption band with v_1 and v_2.

In the infrared spectrum of gaseous NO_2, several other bands are observed[13] as well as the fundamentals. Some of these are clearly overtone and combination bands, but others are characterized by intensities which diminish with increase in temperature. These are therefore attributed to N_2O_4 molecules.

The force constants of the NO_2 molecule have been calculated from its vibrational spectrum.[14,15] For a diatomic molecule, a vibration along the internuclear axis causes displacements, x_1 and x_2, of the two atoms from their equilibrium positions in the same direction on this axis. If the atoms are regarded as point masses and the potential energy function is assumed to be quadratic for small displacements, the potential energy, V, is related to the displacements x_1 and x_2 by the equation

$$2V = F(x_1 - x_2)^2,$$

where F is the force constant of the bond. For a polyatomic molecule, the potential energy function becomes

$$2V = \sum_{ij}^{3N-6} F_{ij}S_iS_j$$

where N = the number of atoms, F_{ij} is the force constant and S_i and S_j are internal displacement coordinates (it being more convenient to use these rather than Cartesian coordinates in the potential energy expression). For NO_2, the potential energy is given by:

$$2V = f_d(\Delta d_1^2 + \Delta d_2^2) + 2f_{dd}\Delta d_1 d_2 + f_\alpha \Delta \alpha^2$$
$$+ 2f_{d\alpha}(\Delta d_1 + \Delta d_2)\Delta \alpha$$

where f_d is the bond-stretching force constant and f_α is the bond-angle deformation force constant and f_{dd} and $f_{d\alpha}$ are interaction force constants which are associated with secondary influences that the nuclei have on each other. The equilibrium bond length is d and the internal coordinates are changes in N,O bond lengths (Δd_1 and Δd_2) and the change in the interbond angle, $\Delta \alpha$. Standard mathematical procedures[16] make it possible to calculate the force constants of a polyatomic molecule from measured vibrational frequencies. In the case of NO_2, there are only three fundamental frequencies and there are four force constants to be calculated. Extra experimental data is therefore needed to make the calculation possible. In addition to the frequencies already mentioned (which relate to $^{14}NO_2$ molecules) it is possible to make use of data on $^{15}NO_2$ provided one assumes that isotopic substitution leaves the force constants unchanged. The force constants determined by Arakawa and Nielsen[15] from the infrared spectra of $^{14}NO_2$ and $^{15}NO_2$ are given in Table III. The constant, f_d, is a measure of the bond strength at internuclear distances close to the equilibrium value d_0 and there is a direct proportionality between f_d and the bond dissociation energy. This is expressed in Morse's law, that f is proportional to d_0^{-6}. The value of f_d for N,O in NO_2 should be compared with f_d for N,N in the nitrogen molecule (2320 Nm^{-1} for a bond length of 0·1075 nm) and for N,N in the N_2O_4 molecule (138 Nm^{-1} for a bond length of 0·1752 nm).[17]

The microwave spectrum of NO_2 is of special interest because it shows hyperfine splitting of the rotational transitions due to interactions involving the umpaired electron of the molecule.

The main isotopic form of nitrogen dioxide, $^{14}N^{16}O_2$, shows[18] five rotational transitions between 15 and 60 GHz, each of which is divided into a multiplet. The spectra of $^{15}NO_2$ and $^{14}N^{16}O^{18}O$ show similar fine structure. In $^{15}NO_2$, rotational states are split as a consequence of two interactions: coupling between the rotational angular momentum, N, and the spin angular momentum, S, of the unpaired electrons; and coupling of the electron spin with the nuclear spin, I. For ^{14}N and ^{15}N, I has the values 1 and $\frac{1}{2}$ respectively, and $S = \frac{1}{2}$.

A typical arrangement of energy levels and transitions in $^{15}NO_2$ is shown in Fig. 4.1.[19] N_2 and N_1 are adjacent rotational levels ($N_2 = N_1 + 1$). Splitting by interaction with electron spin results in two sub-levels, $J = N \pm \frac{1}{2}$. Coupling with nuclear spin leads to further splitting to give the levels, $F = J \pm \frac{1}{2}$. Transitions between different F levels are governed by the selection rule $\Delta F = \Delta J = \Delta N$. There are four 'allowed' transitions, indicated by the full vertical lines in Fig. 4.2. Additional 'forbidden' transitions are possible due to a breakdown in this selection rule. These are shown by dotted lines.

The microwave spectrum of $^{14}NO_2$ is even more complex because further splitting results from interactions involving the nuclear quadrupole moment of ^{14}N as well as

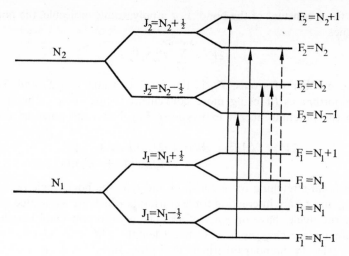

FIG. 4.1. Rotational energy levels and transitions in $^{15}NO_2$.

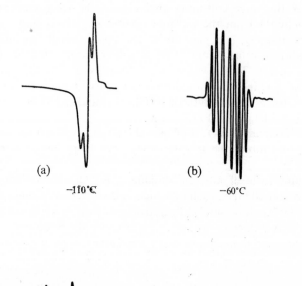

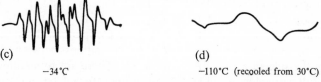

FIG. 4.2. Electron spin resonance spectrum of nitrogen dioxide as a function of temperature (reprinted from *J. Phys. Chem.* **72**, 1721 (1968), copyright 1968 by the American Chemical Society; reprinted by permission of the copyright owner).

those observed in $^{15}NO_2$. Although the analysis of microwave spectroscopic data for $^{14}NO_2$ presents formidable mathematical problems, these have been solved recently[17] and the values for the molecular parameters given in Table III have been obtained.

Structure of Dinitrogen Tetroxide

From an X-ray diffraction study[20] of the solid at $-40°C$, it has been found that the N_2O_4 molecule is centrosymmetrical, all six atoms are co-planar and the two NO_2 groups are joined by a bond between the two nitrogens. The bond lengths and bond angle have been measured at different temperatures[21] (Table IV) and it is found that both the N,N and N,O bonds increase in length with temperature. Notable features of this molecular structure are the unusually long N,N bond, compared, for instance, with the single N,N bond length of 0·147 nm in hydrazine, and the wide O–N–O angle.

TABLE IV

Molecular Parameters of Dinitrogen Tetroxide as determined by X-ray Crystallography and Electron Diffraction

	Cartwright and Robertson[21]			Smith and Hedberg[22]
Temperature (°C)	−248	−128	−13	−20
N,N, bond length (nm)	0·1712	0·1726	0·1745	0·1750
N,O bond length (nm)	0·1189	0·1209	0·1214	0·1180
O–N–O (deg)	134·8 ± 1·5	133·1 ± 0·7	133·7 ± 1·2	133·7

An electron diffraction study[22] of N_2O_4 in the gas phase, measured on a sample cooled to $-20°C$ and so containing very little NO_2, shows very similar molecular dimensions. These results are further confirmed by the structures of the metastable monoclinic form of N_2O_4[23] and the 1:1 molecular compound[24] formed between N_2O_4 and 1,4-dioxane, in both of which the N,N bond length is also found to be 0·175 nm.

The infrared spectrum[25] of solid N_2O_4 is consistent with a planar V_h or a staggered V_d molecular configuration. For both the V_h configuration, in which the two NO_2 groups are co-planar, and the V_d configuration, in which they are perpendicular to each other, the number of Raman and infrared active bands are the same. As the X-ray diffraction data leaves no doubt that V_h is the correct configuration, the infrared spectrum has been interpreted accordingly. The relevant symmetry classes are:

$$3A_g + 2B_{1g} + B_{2g} + A_u + B_{1u} + 2B_{2u} + 2B_{3u}$$

A_g, B_{1g} and B_{2g} are Raman active, B_{1u}, B_{2u} and B_{3u} are infrared active and A_u is inactive in both the Raman and infrared spectra. The five fundamentals active in the infrared have been identified with bands observed at 1728, 1255, 737, 442 and 346 cm^{-1} in the solid at $-180°C$. It is possible to make several sets of assignments for the complete vibrational spectrum of N_2O_4, all of which are quite reasonable, but there appears to be insufficient evidence to prefer one set above all others. Using one possible set, Snyder and Hisatsune[25] have compared the statistical entropy for N_2O_4 with the entropy evaluated from calorimetric studies and hence estimated that the potential barrier to free rotation of the two NO_2 groups is 12·03 kJ mol^{-1}.

The infrared spectrum of N_2O_4 has been studied[26] at low temperatures in matrices of Ar, CO_2, O_2, N_2, N_2O and H_2 as well as in the pure state. Marked differences in the spectrum are found depending on the nature of the sample and the temperature at which it is examined (within the range -269 to $-196\,°C$). The existence of the symmetrical dimer $O_2N \cdot NO_2$ has been amply confirmed in this work, but there is also evidence for the existence of an unsymmetrical dimer, $ONO \cdot NO_2$, in argon matrices at $-269\,°C$. Four infrared bands, located respectively at 1829, 1645, 1290 and 787 cm^{-1}, are believed to characterize this. This could have a planar or staggered configuration. So also can $O_2N \cdot NO_2$, and it is likely that N_2O_4 adopts one of several possible configurations at low temperatures, the particular one found depending on the molecular environment.

E.S.R. Spectrum

Several accounts of the E.S.R. spectrum of nitrogen dioxide have been published.[27-29] The spectrum shows the triplet structure expected for the interaction of the unpaired electron with the nuclear spin, $I = 1$, of ^{14}N. This splits the resonance into $2I + 1$ equally spaced lines. The separation between successive lines observed[27] on dilute solutions of NO_2 in CCl_4 has been reported as $10.76\,mT$ ($\equiv 107.6$ gauss). This is about twice as great as the splitting found for NO_2 trapped in inert matrices or in ice. For example, the splitting observed[28] for NO_2 in ice at $-196\,°C$ is $5.688\,mT$.

The characteristic triplet has been found[29] also in the E.S.R. spectra of NO_2 frozen in ice, nitromethane, chloroform or carbon tetrachloride at temperatures between $-140°$ and $-120°C$. The average splitting of the lines in the triplet is about $5.5\,mT$ in all solvents. When the spectrum is measured on samples at higher temperatures, the triplet is split into more lines, the number and intensities of which vary with the temperature and nature of the matrix. In nitromethane solution, the triplet observed at $-110°C$ is split into nine symmetrical lines separated by $0.53\,mT$ at $-60°C$. These are further split into a 15-line spectrum on warming to $-34°C$. On cooling the sample back to $-110°C$, a broad, unresolved spectrum is observed instead of the original triplet. These changes are illustrated in Fig. 4.2.

Such splitting could arise from the coupling of the spin of the unpaired electron with rotational moments of the NO_2 molecule, from radical–matrix or from radical–radical interactions but since the E.S.R. spectrum is independent of the nature of the matrix it appears more likely to be due to interaction between NO_2 and N_2O_4 molecules present in the matrix. The 15-line spectrum is the result of interaction of the unpaired electron on NO_2 with three nitrogen nuclei, two of which belong to an N_2O_4 molecule. The intermediate 9-line spectrum could then be due to incomplete resolution of the 15-line spectrum at the lower temperatures. This interpretation indicates that a complex of some kind is formed between NO_2 and N_2O_4. The formation of such a complex would also explain the broad spectrum formed on cooling a sample again to $-110°C$. This could correspond to an envelope of unresolved absorption lines.

The E.S.R. spectrum of NO_2 in a polycrystalline matrix of N_2O_4, observed at $-196\,°C$ after ultraviolet irradiation, has been analysed in detail[30-32] and interpreted in terms of the rotation of some of the NO_2 molecules and, at high NO_2 concentrations, interactions between pairs of NO_2 radicals.

Bonding

The bonding within molecules of general formula AB_2 has been discussed by Walsh in a comprehensive series of papers.[33] He has correlated the lowest-energy molecular orbitals which can be constructed from s and p atomic orbitals for a linear molecule with those for a molecule in which the bond angle is 90°. Figure 4.3 shows the correlation proposed by Walsh between the two sets of M.O.s.

For a linear molecule, the orbitals of lowest energy are $1\sigma_g$ and $1\sigma_u$ and are effectively lone-pair orbitals on each B atom which play virtually no part in bonding B to A. Their energy is therefore practically unchanged by a change in bond angle. $2\sigma_g$, $2\sigma_u$ and $1\pi_u$ represent bonding M.O.s., $1\pi_g$ is nonbonding between A and B and $2\pi_u$, $3\sigma_g$ and $3\sigma_u$ are anti-bonding with respect to A and B. On decrease in the bond angle below 180°, the degeneracy of the π-orbitals is removed and additional energy levels are established. Of the eight orbitals of lowest energy on the right-hand side of Fig. 4.3 only one decreases in energy with decrease in bond angle. All the others either increase or are relatively unchanged in energy. A molecule AB_2 which contains 16 or fewer valence electrons should therefore be linear in its ground state.

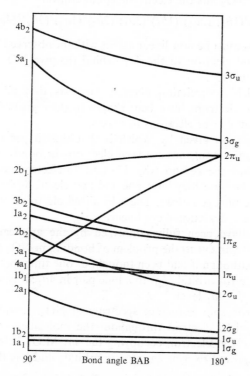

FIG. 4.3. Correlation with bond angle of the energies of the molecular orbitals of a molecule AB_2 (after Walsh, A. D., *J. Chem. Soc.* 2266 (1953)).

Examples of linear molecules and ions which illustrate this are CO_2, COS, CS_2, N_2O, N_3^- and NO_2^+. The ground state of NO_2^+ has the electronic configuration:

$$(1\sigma_g)^2(1\sigma_u)^2(2\sigma_g)^2(2\sigma_u)^2(1\pi_u)^4(1\pi_g)^4$$

In the first excited state of the ion, an electron is excited from the $1\pi_g$ to the $2\pi_u$ orbital. Due to the steep slope of the $4a_1$ component of the latter, the excited NO_2^+ ion is predicted to adopt a bent shape with bond angle around 120°.

45

In the case of the NO_2 molecule, which contains 17 valence electrons, the unpaired electron will be in the $4a_1$ orbital and so a bent molecule will be of lower energy than a linear one. The $4a_1$ orbital may be described approximately as an s orbital localized on the nitrogen atom. The other orbitals below are described thus:

$1a_1$ and $1b_2$ (lone-pairs);
$2a_1$ (bonding);
$2b_2$ (bonding, partly σ and partly π);
$1b_1$ (bonding);
$3a_1$ (largely bonding);
$3b_2$ (non-bonding);
$1a_2$ (non-bonding).

These are filled by the remaining 16 valence electrons from nitrogen and the two oxygens and the ground state of NO_2 has the electronic configuration:

$$(1a_1)^2(1b_2)^2(2a_1)^2(1b_1)^2(3a_1)^2(2b_2)^2(1a_2)^2(3b_2)^2(4a_1).$$

The conclusion that NO_2 must be non-linear agrees with the observed bond angle of $134.25°$. The addition of a second electron to the $4a_1$ orbital (to give NO_2^-) produces a further decrease in bond angle to $115°$.

More generally, Walsh's correlation diagram suggests that all AB_2 molecules with 17, 18, 19 or 20 valence electrons have bent shapes in their ground states and the bond angle decreases as the number of electrons increases.

An alternative explanation given by Walsh is that the shapes of AB_2 molecules are determined by electrostatic repulsions between electrons in bond orbitals and those in a "lone-pair" orbital. Thus the reason why 17- and 18-electron molecules have ground states which are bent is because they have one and two electrons respectively which are in orbitals localized on the nitrogen atom. These localized electrons repel the bonding pairs of electrons and so force a deviation from linearity.

Walsh's conclusions have been fully confirmed by the application[34-36] of Hückel's method using a LCAO treatment to the problem of bonding in NO_2 for bond angles within the range 90° to 180°. Although it had been thought at one time that the unpaired electron should be assigned to the $1a_2$ orbital,[37] with a lone pair in $4a_1$, it is now generally accepted as correct to place it in the $4a_1$ level.

Analysis of the electron spin resonance spectrum of NO_2 provides information of the distribution of the unpaired electron throughout the molecule. The spectrum has been interpreted[28] in terms of this electron occupying the $4a_1$ M.O. This is largely constructed from nitrogen $2s$ and $2p_z$ and oxygen $2p_z$ atomic orbitals and has the form:

$$\psi(4a_1) = a_s\psi_N(2s) + a_z\psi_N(2p_z) + b_1\psi_O(2p_z)$$

where the z axis is the C_{2v} symmetry axis, the x axis is perpendicular to the molecular plane and the y axis is parallel to the O,O direction. A representative set of results[31] shows that the unpaired electron density is distributed in the various orbitals in the following way: s, 0.097; $2p_z$, 0.371; and $2p_z$ (oxygens), 0.53.

Comparative studies of the E.S.R. spectra of NO_2 and the isoelectronic ion, CO_2^-, have shown that the unpaired electron in NO_2 is delocalized much more than that in CO_2^-. This greater delocalization means that dimerization is more difficult for NO_2 than for CO_2^- so that, despite the charge repulsion between two CO_2^- radicals, the oxalate ion is stable whereas N_2O_4 readily dissociates.

The chief point of interest in the structure of the dimer, $O_2N \cdot NO_2$, is the long and weak N,N bond. Other unexpected properties are the largeness of the ONO angle, the planarity of the molecule and the barrier to rotation about the central bond. When these structural features are considered together, it becomes very difficult to provide a really satisfactory theoretical account of the bonding in the dimer.

This is illustrated by the various descriptions proposed from time to time for the N,N bond. It has been represented as largely of σ-type, involving the $4a_1$ orbitals of the NO_2 fragments,[36] and a small amount of π-bonding originating from the $1b_1$ orbitals of NO_2. The converse view[35] is that the $4a_1$ orbitals contribute very little to the bonding between the nitrogens. Alternatively, the properties of the bond have been explained in terms of the delocalization of $2p\pi$ electrons (originating as lone-pairs on the oxygen atoms) into an antibonding σ-orbital.[38]

Other suggestions include a 'π-only bond'[37] and a 'splayed' single bond arising from electron correlation effects.[39] It appears to be generally accepted that the nitrogens are joined by a σ- rather than a π-bond. The nature of the orbitals involved remains in doubt and a full interpretation of the structure of N_2O_4 is still awaited.

References

1. MELDRUM, A., *J. Chem. Soc.* 902 (1933).
2. GIAUQUE, W. F., and KEMP, J. D., *J. Chem. Phys.* **6**, 40 (1938).
3. STEESE, C. M., and WHITTAKER, A. G., *J. Chem. Phys.* **24**, 776 (1956).
4. HARRIS, L., and CHURNEY, K. L., *J. Chem. Phys.* **47**, 1703 (1967).
5. HISATSUNE, I. C., *J. Phys. Chem.* **65**, 2249 (1961).
6. REDMOND, T. F., and WAYLAND, B., *J. Phys. Chem.* **72**, 1626 (1968).
7. CLUSIUS, K., and VECCHI, M., *Helv. Chim. Acta* **36**, 930 (1953).
8. MILLEN, D. J., *J. Chem. Soc.* 2600 (1950).
9. MOORE, G. E., *J. Opt. Soc. Amer.* **43**, 1045 (1953).
10. BIRD, G. R., *J. Chem. Phys.* **25**, 1040 (1956).
11. CLAESSON, S., DONOHUE, J., and SCHOMAKER, V., *J. Chem. Phys.* **16**, 207 (1948).
12. HARRIS, L., BENEDICT, W. S., and KING, G. W., *Nature* **131**, 621 (1933).
13. WILSON, M. K., and BADGER, R. M., *Phys. Rev.* **76**, 472 (1949).
14. WESTON, R. E., *J. Chem. Phys.* **26**, 1248 (1957).
15. ARAKAWA, E. T., and NIELSEN, A. H., *J. Mol. Spectry.* **2**, 413 (1958).
16. WALKER, S., and STRAW, H., *Spectroscopy*, Vol. II. Chapman Hall, 1962.
17. DECIUS, J. C., *J. Chem. Phys.* **45**, 1069 (1966).
18. BIRD, G. R., *J. Chem. Phys.* **25**, 1040 (1956).
19. BIRD, G. R., BAIRD, J., JACHE, A. W., HODGESON, J. A., CURL, R. F., ANDERSEN, J. R., ROSENTHAL, J., KUNKLE, A. C., and BRANSFORD, J. W., *J. Chem. Phys.* **40**, 3378 (1964).
20. BROADLEY, J. S., and ROBERTSON, J. M., *Nature* **164**, 915 (1949).
21. CARTWRIGHT, B. S., and ROBERTSON, J. M., *Chem. Comm.* 82 (1966).
22. SMITH, D. W., and HEDBERG, K., *J. Chem. Phys.* **35**, 1282 (1956).
23. GROTH, P., *Nature* **198**, 1081 (1963).
24. GROTH, P., and HASSEL, O., *Proc. Chem. Soc.* 379 (1962).
25. SNYDER, R. G., and HISATSUNE, I. C., *J. Mol. Spectry.* **1**, 139 (1957).
26. FATELY, W. G., BENT, H. A., and CRAWFORD, B., *J. Chem. Phys.* **31**, 204 (1959).
27. BIRD, G. R., BAIRD, J. C., and WILLIAMS, R. B., *J. Chem. Phys.* **28**, 738 (1958).
28. ATKINS, P. W., KEEN, N., and SYMONS, M. R. C., *J. Chem. Soc.* 2873 (1962).
29. BIELSKI, B. H. J., FREEMAN, J. J., and GEBICKI, J. M., *J. Phys. Chem.* **72**, 1721 (1968).
30. SCHAAFSMA, T. J., VAN DER VELDE, G., and KOMMANDEUR, J., *Mol. Phys.* **14**, 501 (1968).
31. SCHAAFSMA, T. J., and KOMMANDEUR, J., *Mol. Phys.* **14**, 517 (1968).
32. SCHAAFSMA, T. J., and KOMMANDEUR, J., *Mol. Phys.* **14**, 525 (1968).
33. WALSH, A. D., *J. Chem. Soc.* 2266 (1953).

34. KATO, H., YONEZAWA, T., MOROKUMA, K., and FUKUI, K., *Bull. Chem. Soc. Jap.* **37**, 1710 (1964).
35. GREEN, M., and LINNETT, J. W., *Trans. Faraday Soc.* **57**, 1 (1961).
36. BURNELLE, L., BEAUDOUIN, P., and SCHAAD, L. J., *J. Phys. Chem.* **71**, 2240 (1967).
37. COULSON, C. A., and DUCHESNE, J., *Bull. Acad. Belg. Cl. Sci.* **43**, 522 (1957).
38. BROWN, R. D., and HARCOURT, R. D., *Aust. J. Chem.* **16**, 737 (1963).
39. BENT, H. A., *Inorg. Chem.* **2**, 747 (1963).

CHAPTER 5

TRISILYLAMINE

Introduction

The representative elements of Group V form the simple hydrides of general formula XH_3 (X = N, P, As or Sb). Ammonia and its derivatives have been extensively studied and have many important uses. The other hydrides are less stable and their chemistry is not so diverse, but their derivatives, wherein one or more of the hydrogens are replaced by other atoms or groups, are better known.

The valence shell of a Group V atom has the electron configuration ns^2np^3 in the isolated state. In the combined state in the hydride, it contains three bonding pairs and one lone pair of electrons. The chemical properties and molecular shape of the hydride are both strongly influenced by the presence of this lone pair.

Thus a characteristic property of ammonia and substituted ammonias, and their phosphine, arsine and stibine analogues, is their behaviour as Lewis bases. This is due to the lone pair on the central atom which can be donated to acceptor molecules or ions. The tendency to show donor properties depends, *inter alia*, on the electronegativity of the other atoms or groups combined with the central atom.

The molecules NH_3, PH_3 and AsH_3 and their corresponding trimethyl derivatives are all known to be pyramidal. This has been conventionally described in terms of sp^3 hybridization of the four orbitals in the valence shell of the Group V atom. These hybrids are tetrahedrally directed and the pyramidal shape of the XH_3 molecule results from the location of the lone pair in one of these and the bonding pairs in the other three. The bond angle is often significantly smaller than the tetrahedral value (109·5°) and this is interpreted in terms of repulsion between a lone pair and a bonding pair being greater than that between two bonding pairs, with a consequent closing-up of the bond angle.

Trisilylamine, $N(SiH_3)_3$, is formally derived from ammonia by the replacement of three hydrogens with silyl groups. This compound was first synthesized in 1921 by Stock and Somieski,[1] but until quite recently remained the sole representative of amines triply substituted by silyl groups. This is in sharp contrast to the wide range of triply substituted organic derivatives of ammonia which has been known for over a century. In the past few years there have been synthesized[2] well over 100 compounds which contain the atomic grouping NSi_3, the majority of these containing alkyl, alkoxy, dialkylamine or halogen as substituents on the silicon atom. Quaternary compounds analogous to substituted ammonium compounds are not known. This is another point of major difference and shows that the donor properties of trisilylamine and its derivatives must be very weak. This absence of significant basic character has been related to the participation of the lone pair on nitrogen in $(p \rightarrow d)$ π-bonding to silicon so that it is no longer available for inter-action with a Lewis acid.

Preparation

Trisilylamine was first prepared[1] by mixing ammonia and excess silyl chloride in the vapour phase

$$3SiH_3Cl + 4NH_3 \rightarrow N(SiH_3)_3 + 3NH_4Cl$$

Silyl chloride can be conveniently prepared from silicon tetrachloride by reduction to silane in diethylether solution using lithium aluminium hydride, and then chlorination of the silane with hydrogen chloride and aluminium chloride to produce SiH_3Cl.

It has also been made[3] by combination between ammonia and liquid silyl bromide at room temperature.

The reaction of a halogen-substituted silane with ammonia proceeds according to the following mechanism:

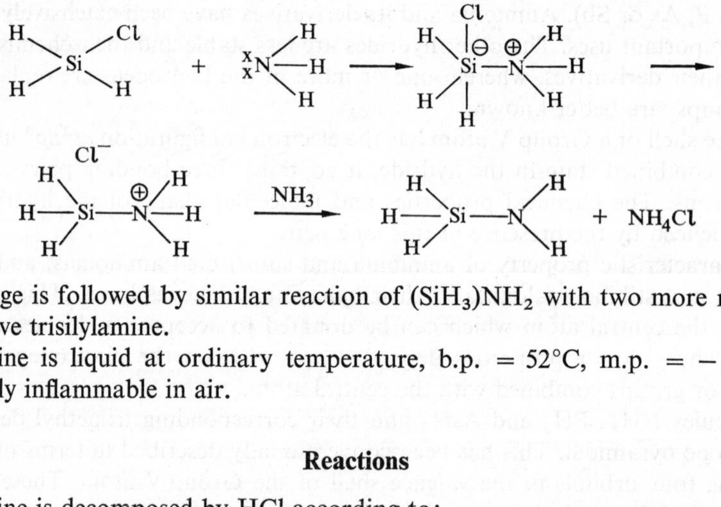

This first stage is followed by similar reaction of $(SiH_3)NH_2$ with two more molecules of SiH_3Cl to give trisilylamine.

Trisilylamine is liquid at ordinary temperature, b.p. = 52°C, m.p. = −105°C. It is spontaneously inflammable in air.

Reactions

Trisilylamine is decomposed by HCl according to:

$$N(SiH_3)_3 + 4HCl \rightarrow 3SiH_3Cl + NH_4Cl$$

Its extreme weakness as an electron pair donor is shown by the difficulty in isolating adducts with electron acceptor molecules. For example, $N(SiH_3)_3BF_3$ can be formed[4] only at low temperatures (between −40 and −78°C) and decomposes irreversibly above −40°C. It is possible that an addition compound with BCl_3 can be formed at −78°C, but this decomposes slowly even at this low temperature so that the overall reaction is

$$N(SiH_3)_3 + BCl_3 \rightarrow SiH_3Cl + (SiH_3)_2NBCl_2$$

In contrast, the adducts between BF_3 or BCl_3 and $(CH_3)_3N$ are much more stable. Tri-methylamine also forms a stable addition compound, $(CH_3)_3N \cdot BH_3$, by reaction with diborane, but trisilylamine undergoes no parallel reaction.

50

There is a systematic decrease in the stability of addition compounds when carbon atoms are successively replaced by silicon in the series: $(CH_3)_3N$, $(SiH_3)(CH_3)_2N$, $(SiH_3)_2 \cdot CH_3N$ and $(SiH_3)_3N$. Only the first two compounds form addition compounds[3] with trimethylboron, that with N-silyldimethylamine being the less stable. Further, both B_2H_6 and $[(CH_3)_3Al]_2$ form adducts with the first three compounds, but only the very strong Lewis acid, $[(CH_3)_3Al]_2$, combines with $(SiH_3)_3N$. The order of basicity suggested by these results is:

$$(CH_3)_3N > (SiH_3)(CH_3)_2N > (CH_3)(SiH_3)_2N > (SiH_3)_3N$$

In resonance theory, the lack of basic character of trisilylamine is attributed to contributions by structures involving double bonds:

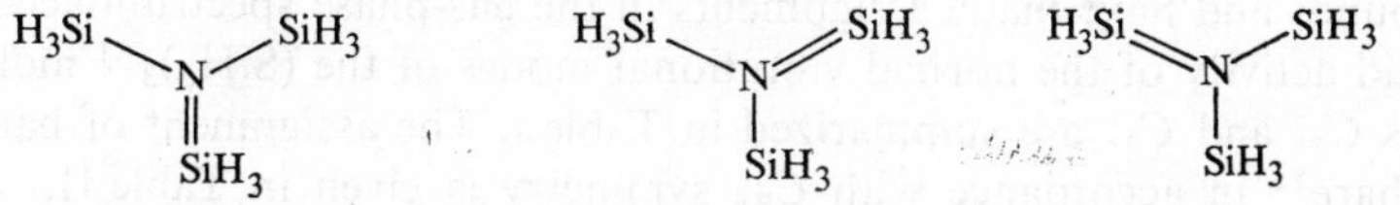

If these are important contributions, then the lone pair electrons on nitrogen are less available for coordination with acceptor molecules. The participation of such double-bonded structures would necessitate the expansion of the valence shell of silicon from 8 to 10 electrons.

Structure

A study of trisilylamine by electron diffraction[5] has shown that the silicon and nitrogen atoms are coplanar. The molecular parameters are: Si,N bond length = $0 \cdot 1738 \pm 0 \cdot 002$ nm Si,H = $0 \cdot 154 \pm 0 \cdot 005$ nm; Si,Si non-bonded distance = $0 \cdot 3005 \pm 0 \cdot 002$ nm; and the average Si–N–Si bond angle = $120°$. The Si,N bond length is shorter than that expected for a single bond, estimated values for which are $0 \cdot 180^6$ and $0 \cdot 187^7$ nm. Corresponding estimates of Si,N double bond length are $0 \cdot 162^6$ and $0 \cdot 167^7$ nm and it appears that the Si,N bonds in trisilylamine have partial double bond character.

Further confirmation of the planarity of the Si_3N skeleton comes from the zero dipole moment of trisilylamine in the gas phase[8] between $21°$ and $90°C$. This can only be interpreted in accordance with a centrosymmetric (i.e. planar) grouping of Si_3N.

Vibrational Spectrum

Among studies of this are measurements of the infrared spectrum in the gas phase,[9,10] Raman spectrum as a liquid[9,12] at $0°C$, and the infrared spectrum of the solid at $-196°C$ and of its solution in an argon matrix at $-253°C$.[11] The assignment of some of the fundamental vibrations in the spectrum of gaseous samples cannot be made with certainty because the absorption bands are fairly broad and there is considerable overlap between them. There is incomplete resolution of some bands even in the spectrum of the argon matrix, although they are sharper and observable in more detail than on a gaseous sample.

The spectra can be accordingly interpreted equally well using molecular models of different symmetries. These are all based on a planar skeleton for the Si_3N grouping, but differ in the configurations of hydrogen atoms in the silyl groups. If the silyl groups undergo free rotation, the effective point group of the molecule would be D_{3h}. This is composed of the following main symmetry elements: a principal C_3 axis perpendicular to the Si_3N

plane, three C_2 axes perpendicular to C_3, and the symmetry plane, σ_h, containing Si_3N. If, on the other hand, there are preferred configurations for the silyl group, then different, lower symmetries such as C_{3h}, C_{3v}, C_3, C_s or C_1, would be possible. Low symmetries like C_3, C_s or C_1 are considered unlikely and it is probable that, if there is $(p \rightarrow d)\pi$-bonding between N and Si, the silyl groups cannot rotate freely. The most common interpretations of the vibrational spectrum have therefore been in terms of C_{3h} or C_{3v} symmetry.

The orientations of the silyl groups for the symmetries C_{3h} and C_{3v} according to Ebsworth and co-workers are shown in Fig. 5.1. For C_{3h} symmetry, one hydrogen of each SiH_3 group is coplanar with Si_3N and the remaining hydrogens are above and below this plane respectively. For C_{3v} symmetry, one hydrogen is above and the other two are below the Si_3N grouping. Ebsworth and co-workers have concluded that C_{3h} is more plausible on steric grounds and have made assignments in the gas-phase spectrum accordingly. The symmetry and activity of the normal vibrational modes of the $(SiH_3)_3N$ molecule for the point groups C_{3h} and C_{3v} are summarized in Table I. The assignment of bands by Goldfarb and Khare[11] in accordance with C_{3v} symmetry is given in Table II.

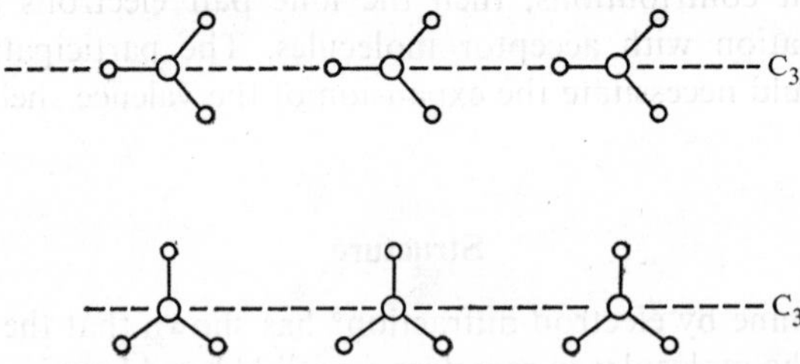

FIG. 5.1. Orientations of silyl groups in trisilylamine for the symmetries C_{3h} and C_{3v} (following Ebsworth, E. A. V., et al., Spectrochim. Acta 13, 202 (1958)). The circle upon which the three silicon atoms lie is, for clarity, shown opened out as a dotted line.

TABLE I

Symmetry and Activity of the Vibration Modes of $N(SiH_3)_3$ for the Point Groups C_{3h} and C_{3v}

Point group C_{3h}

	A'(R)	A''(i.r.)	E'(i.r. + R)	E''(R)
Si–H stretch	ν_1, ν_2	ν_7	ν_{12}, ν_{13}	ν_{19}
SiH$_3$ deform.	ν_3, ν_4	ν_8	ν_{14}, ν_{15}	ν_{20}
SiH$_3$ rock	ν_5	ν_9	ν_{16}	ν_{21}
N–Si stretch	ν_6		ν_{17}	
N–Si$_3$ deform.		ν_{10}	ν_{18}	
SiH$_3$ torsion		ν_{11}		ν_{22}

Point group C_{3v}

	A_1(i.r. + R)	A_2 (inactive)	E (i.r. + R)
Si–H stretch	ν_1, ν_2	ν_8	$\nu_{12}, \nu_{13}, \nu_{14}$
SiH$_3$ deform.	ν_4, ν_3	ν_9	$\nu_{15}, \nu_{16}, \nu_{17}$
SiH$_3$ rock	ν_5	ν_{10}	ν_{18}, ν_{19}
N–Si stretch	ν_6		ν_{20}
N–Si$_3$ deform.	ν_7		ν_{21}
SiH$_3$ torsion		ν_{11}	ν_{22}

TABLE II
Assignment by Goldfarb and Khare of Bands in the Argon-Matrix Spectrum of $N(SiH_3)_3$ on the Assumption of C_{3v} Symmetry

Frequency (cm^{-1})	Vibration assigned
2193	ν_{12}
2186	ν_{13}
2174	ν_1
2155	ν_2, ν_{14}
1192	$\nu_6 + \nu_{19}$
1115	$\nu_4 + \nu_{21}$
1007	ν_{17}
998	$2\nu_6$
966	ν_{15}
955	ν_{16}
942	ν_3
934	ν_{17}
921	ν_4
767	ν_5
743	ν_{18}
693	ν_{19}
454	ν_7

It is possible that the trisilylamine molecule changes from a C_{3h} to a C_{3v} structure upon condensation in the argon matrix so that an interpretation on either basis, referring as it does to a different phase, is correct. It does not appear possible, in view of the overlapping observed between certain bands, to reach an unequivocal decision between the two possible models with the experimental evidence available at present. The vibrational spectrum does strongly favour a planar, and not a pyramidal, Si_3N skeleton. For example, there is a parallelism between bands assigned to this skeleton and the fundamental vibrations of the planar BCl_3 molecule.

The Raman spectra of liquid trisilylamine, observed by Ebsworth and co-workers[9] and by Kriegsmann and Förster,[12] both contain fewer lines than the number expected for either C_{3h} or C_{3v} symmetry and so do not assist in deciding between these alternatives. There are significant differences between the two sets of Raman frequencies. $\Delta\nu$ (cm^{-1}) values reported by Ebsworth are 204 (dp), 496 (p), 946 (dp), 987 (dp) and 1011 (dp). Those given by Kriegsmann and Förster are 248, 482 and 524 (assigned as vibrations of the Si_3N framework), 581 (SiH_3 rocking), 892 and 949 (antisymmetric deformations of SiH_3) and 2180, 2200 and 2288 (Si–H stretches). The differences may be at least partly caused by experimental difficulties in measuring the spectrum due to the decomposition of the compound which is accelerated by the radiation used to excite the Raman effect.

It is reasonable to suppose that the vibrations of the Si_3N skeleton are little affected by the configuration of the hydrogen atoms and so the symmetry of this portion of the molecule may be considered separately from that of the rest. Four normal vibrations of the Si_3N are thus to be expected, that is, those which characterize a tetratomic planar molecule (Fig. 11.3, p. 106).

Such a shape should give two depolarized Raman lines, ν_{17} and ν_{18}, associated with stretching and bending modes respectively and each coincident with an infrared active band; one polarized Raman line, ν_6, due to symmetric stretching and having no coincident band in the infrared; and one infrared band, ν_{10}, due to out-of-plane deformation, with no Raman counterpart. In contrast, a pyramidal Si_3N skeleton should give two polarized and two depolarized Raman lines, all with coincidences in the infrared. Ebsworth and co-workers observed only one polarized Raman line (at 496 cm^{-1} and identified with ν_6) in the region below 1000 cm^{-1}, where skeletal frequencies are expected to occur. No infrared band due to ν_7 was observed but probably this would be located below 400 cm^{-1}, the lower limit of their frequency measurements. A second Raman line, observed at 204 cm^{-1}, was attributed, by analogy with the spectrum of BCl_3, to the in-plane deformation of the Si_3N grouping (ν_{18}). The infrared band found at 996 and the Raman line at 987 cm^{-1} were identified with the asymmetric stretch, ν_{17}.

N.M.R. Spectrum

The proton resonance spectrum of $^{14}N(^{28}SiH_3)_3$ shows[13] a single resonance at 40 MHz. This is broadened because of quadrupole coupling of the proton resonance with the ^{14}N nuclei ($I = 1$). In $^{15}N(^{28}SiH_3)_3$, the resonance is split into two equal components due to coupling with the central nitrogen nucleus of spin $\frac{1}{2}$.

In $H_3^{29}Si^{15}N(^{28}SiH'_3)_2$, the hydrogens bound to ^{28}Si ($I = 0$) are magnetically non-equivalent to those bound to ^{29}Si ($I = \frac{1}{2}$). In the pure liquid, the coupling constant $J(H^{29}Si)$ for the directly bonded proton is $213\cdot8 \pm 0\cdot2$ Hz. $J(H'^{29}Si)$, for the distantly bonded protons, is $3\cdot74 \pm 0\cdot05$ Hz and so there is effectively a large chemical shift between the two sets of protons. $J(HH')$ is $0\cdot37 \pm 0\cdot1$ Hz.

It has been observed[14] that the proton chemical shifts in silyl compounds are much less sensitive to the nature of substituents than they are in analogous carbon compounds. For instance, the chemical shifts (τ) relative to cyclohexane ($\tau = 8\cdot56$ ppm) measured on a number of silyl compounds at 40 MHz are:

$$SiH_3F, 5\cdot24; \quad (SiH_3)_2O, 5\cdot39; \quad SiH_3Cl, 5\cdot41;$$
$$(SiH_3)_3N, 5\cdot56; \quad (SiH_3)_2S, 5\cdot65; \quad SiH_3I, 6\cdot56.$$

There is a general decrease in τ as the electronegativity of the substituent increases. The significance of this effect is difficult to assess because the chemical shifts of protons attached to silicon in these compounds is likely to be affected by the $(p \rightarrow d)\pi$-bonding which is likely to occur.

Bonding

The two main features of the structure of trisilylamine which require explanation are the planarity of the NSi_3 grouping and the intermediate length of the Si,N bond between values expected for a single and for a double bond. These are usually discussed in relation to $(p \rightarrow d)\pi$-bonding from nitrogen to silicon. The lone pair in a p-orbital on nitrogen can overlap with a vacant d-orbital of appropriate symmetry on silicon to form the π-bond

FIG. 5.2. Overlap between orbital on nitrogen and silicon to form a $(p \rightarrow d)$ π-bond.

(Fig. 5.2). The prerequisite for the establishment of such bonds is the near or exact co-planarity of the silicons and nitrogen because this arrangement facilitates maximum overlap between the orbitals concerned. This extra bonding stabilizes the co-planar structure. Where it cannot occur, for example, in trimethylamine (the orbitals of appropriate symmetry and energy on the carbon atoms being absent), the pyramidal arrangement of bonds around nitrogen is found.

Ebsworth[15] has, however, pointed out that the co-planarity condition no longer applies if we consider the lone pair to be in a tetrahedral (sp^3) hybrid orbital rather than a pure p-orbital on the nitrogen. In that case, substantial overlap with silicon d-orbitals may be possible with a pyramidal arrangement of bonds. In other words, the stereochemistry around nitrogen is not necessarily a reliable guide to the extent of $(p \rightarrow d)\,\pi$-interaction.

Perkins[16] has calculated, for a planar Si_3N grouping, the extent to which the nitrogen lone pair electrons are delocalized into the d-orbitals of the silicon atoms and estimated that significant $N-Si(p \rightarrow d)\,\pi$-bonding exists with each Si–N bond of mean energy approximately 67 kJ mol^{-1}. The situation is more complicated if the possibility of distortion of the planar Si_3N skeleton to give a pyramidal structure is considered. Then the distinction between σ- and π-molecular orbitals disappears and their respective contributions to the bonding can no longer be differentiated.

A quantitative estimate of the strength of $(p \rightarrow d)\,\pi$-bonding should be available by calculation of the stretching force constant for the Si–N bond. From their vibrational assignments, Ebsworth and co-workers[9] calculated that the effective force constant for symmetric stretching is approximately 440 Nm^{-1}. From a recent study[17] of the vibrational spectrum of the related compound trisilylphosphine, $P(SiH_3)_3$, it has been concluded that the PSi_3 skeleton is planar and that the force constant of the Si–P bond is about 310 Nm^{-1}. Both these values are within what is commonly regarded as the range for single bonds and suggest that the degree of π-bonding cannot be very large, although it may be enough to influence the molecular stereochemistry.

References

1. STOCK, A., and SOMIESKI, K., *Chem. Ber.* **54**, 740 (1921).
2. WANNAGAT, A., *Fortschr. Chem. Forsch.* **9**, 102 (1967).
3. SUJISHI, S., and WITZ, S., *J. Amer. Chem. Soc.* **76**, 4631 (1954).
4. BURG, A. B., and KULJIAN, E. S., *J. Amer. Chem. Soc.* **72**, 3103 (1950).
5. HEDBERG, K., *J. Amer. Chem. Soc.* **77**, 6491 (1955).
6. SCHOMAKER, V., and STEVENSON, D. P., *J. Amer. Chem. Soc.* **63**, 37 (1941).
7. PAULING, L., *Nature of the Chemical Bond*, 3rd ed., Cornell Univ. Press. Ithaca, N.Y., 1960.
8. VARMA, R., MACDIARMID, A. G., and MILLER, J. G., *J. Chem. Phys.* **39**, 3157 (1963).
9. EBSWORTH, E. A. V., HALL, J. R., MACKILLOP, M. J., MCKEAN, D. C., SHEPPARD, N., and WOODWARD, L. A., *Spectrochim. Acta* **13**, 202 (1958).
10. ROBINSON, D. W., *J. Amer. Chem. Soc.* **80**, 5924 (1958).
11. GOLDFARB, T. D., and KHARE, B. N., *J. Chem. Phys.* **46**, 3379 (1967).
12. KRIEGSMANN, H., and FÖRSTER, W., *Zeit. anorg. allgem. Chem.* **298**, 212 (1959).
13. EBSWORTH, E. A. V., and SHELDROCK, G. M., *Trans. Faraday Soc.* **62**, 3282 (1966).
14. EBSWORTH, E. A. V., and TURNER, J. J., *J. Phys. Chem.* **67**, 805 (1963).
15. EBSWORTH, E. A. V., *Chem. Comm.* 530 (1966)
16. PERKINS, P. G., *Chem. Comm.* 268 (1967).
17. DAVIDSON, G., EBSWORTH, E. A. V., SHELDRICK, G. M., and WOODWARD, L. A., *Spectrochim. Acta* **22**, 67 (1966).

CHAPTER 6

NITROUS ACID

Introduction

In 1774, C. W. Scheele noticed the change in properties of potassium nitrate after it has been melted and kept at red heat for 30 minutes. This was related to a loss of oxygen on heating and Berzelius showed that the potassium salts of two nitrogen-containing acids, nitrous and nitric, were involved. Nitrous acid forms nitrites on reaction with bases and these have quite different properties from nitrates.

Although nitrous acid and its salts have been known for many years, the acid has proved difficult to characterize fully because it cannot be made in the pure state. It is invariably found with the compounds from which it is made or with its decomposition products. The presence of these impurities must be fully recognized in order to ensure correct determination of the physical and chemical properties of nitrous acid. There is now substantial evidence which indicates the existence of two isomeric forms of nitrous acid and these have been the subject of much recent research.

Preparation

Nitrous acid is present in the blue solution formed when dinitrogen trioxide, N_2O_3, is dissolved in cold water. A mixture of nitrous and nitric acids is obtained when nitrogen dioxide is dissolved in water.

Aqueous nitrous acid is readily made by adding a mineral acid to the solution of a nitrite:

$$NaNO_2 + HCl \rightarrow NaCl + HONO$$

The free acid in aqueous solution can be conveniently made by an ion-exchange procedure.[1] A solution of sodium nitrite in a mixture of water and ethyleneglycol dimethylether is passed through a column of cation exchange resin in the acid form at $-30°C$. Titration of the column effluent with alkali has shown that at least 97% of the theoretical acidity can be recovered under these conditions. It is necessary to operate at low temperatures and concentrations to reduce the decomposition of nitrous acid to a minimum.

Nitrous acid is formed in the gas phase by reaction between NO, NO_2 and H_2O.

$$NO + NO_2 + H_2O \overset{K_1}{\rightleftharpoons} 2HONO$$

This is a very rapid reaction: at 24°C, the time for half reaction is approximately 14 msec. Its progress has been followed[2] by determination of the concentration of NO_2 from absorbance measurements and the kinetics found to follow a third-order rate law.

The equilibrium constant, K_1, has been calculated[3] to be $0.0136 \text{ kN}^{-1}\text{m}^2$ at 25°C and $0.015 \text{ kN}^{-1}\text{m}^2$ at 19.95°C.

Stability in Aqueous Solution

In aqueous solution, nitrous acid decomposes thus:

$$3HONO \rightleftharpoons 2NO + HONO_2 + H_2O$$

The kinetics of decomposition are consistent with the following reactions:

$$4HONO \rightleftharpoons 2N_2O_3 + 2H_2O$$
$$2N_2O_3 \rightleftharpoons N_2O_4 + 2NO$$
$$N_2O_4 + H_2O \rightleftharpoons HONO + HONO_2$$

As the decomposition leads to evolution of nitric oxide, the rate is influenced by stirring the solution or by the addition of certain catalysts.

In dilute solution, nitrous acid exists mainly as HONO molecules. The above reactions show that the formation of N_2O_3 is promoted in concentrated solutions, that is, when the activity of water is low. Hence the decomposition rate increases with concentration.

Reactions

Compounds of general formula NOX (X = Cl, Br, HSO_4 or NO_3) are formed in aqueous solution.

$$HONO + HX \rightleftharpoons NOX + H_2O$$

Nitrous acid is oxidized to nitric acid by ozone, hydrogen peroxide and other oxidizing agents. Reducing agents like iodide or iron(II) salts convert it to nitric oxide.

Nitrous acid is widely used as a reagent in synthetic organic chemistry. With certain classes of compound, nitrous acid reacts to form N- or C-nitroso-derivatives. For example, the nitrosation of aliphatic ketones, $RCO \cdot CH_2 \cdot R$, proceeds via attack by NO^+, to give $R \cdot CO \cdot CR \cdot NOH$. Primary aliphatic amines are determined:

$$RNH_2 + HONO \rightleftharpoons ROH + H_2O + N_2$$

Reaction with aromatic amines gives diazonium compounds.

$$2HONO \rightleftharpoons N_2O_3 + H_2O$$
$$RNH_2 + N_2O_3 \rightarrow R \cdot NH \cdot NO + HONO$$
$$RNH \cdot NO + HX \rightarrow RN_2X + H_2O$$

C-nitroso-derivatives are obtained by reaction between nitrous acid and aromatic compounds such as phenols.

Nitrous Acid Molecules

The absorption spectrum of a gaseous mixture of NO, NO_2 and H_2O or D_2O shows bands[4] in the region 310–390 nm which can only be due to nitrous acid molecules. The presence of water vapour is necessary for the appearance of these bands and they show a pronounced wavelength shift when H_2O is replaced by D_2O. Nitric acid molecules are also likely to be present in equilibrium mixtures but, as this compound gives a continuous spectrum in this wavelength region, it can be easily differentiated from nitrous acid.

The presence of nitrous acid molecules in aqueous solutions of N_2O_3 is shown by similar absorption in the ultraviolet.[5] Another species is present which absorbs between 300 and 400 nm. This second species does not show the vibrational fine structure which characterizes the absorption of nitrous acid in the same region and it has been tentatively identified as $(HNO_2NO)^+$.

Equilibria involving HONO and NO^+ in the presence of a strong acid have been studied.[6] Molecular HONO exists at low acidities but at higher acidities NO^+ is formed and the characteristic ultraviolet absorption spectrum of HONO is no longer observed. In a mixture of molar composition $H_2SO_4:HONO:H_2O = 43:10:47$, the nitrous acid is almost completely converted to NO^+.

Possibly $H_2NO_2^+$ ions could also be formed, by proton transfer from the strong acid, but there appears to be little evidence to support the formation of these in any significant quantity.

In 10% perchloric acid[7] nitrous acid molecules are present, but in 58% acid these are all converted to NO^+.

The concentrations of HONO and NO^+ in acidic media have been measured spectrophotometrically[8] and the equilibrium constant, K_2, for the reaction

$$HONO + H^+ \overset{K_2}{\rightleftharpoons} H_2O + NO^+$$

determined.

$$K_2 = \frac{\{NO^+\} \cdot \{H_2O\}}{\{HONO\} \cdot \{H^+\}}$$

and values found were 2.6×10^{-5} in aqueous sulphuric acid at $4.5°C$, and 8.7×10^{-7} in aqueous perchloric acid at $3.5°C$. Accurate measurements in acidic media are hindered by the fact that in the range of mineral acid concentration $(40-57\%)$ where HONO and NO^+ are both present in significant proportions, the decomposition rate of nitrous acid is highest.

Dissociation Constant of Nitrous Acid

The ionization of nitrous acid has been studied by spectrophotometry, conductometry and potentiometry. The results of some of the early investigations are not very reliable because of the partial decomposition of nitrous acid that must have occurred during the measurements. Thus it is known that the decomposition of a solution of sodium nitrite containing 0.01md dm^{-3} is catalysed[9] by the presence of a small concentration of a mineral acid or even a weak organic acid like acetic. The presence of sodium nitrate inhibits decomposition by acetic acid. The most accurate values for the dissociation constant of nitrous acid are those[10] determined by potentiometric titration of sodium nitrite solutions, in the presence of nitrate, with acetic acid as titrant. Table I shows the pK values for a range of temperature and the thermodynamic values, pK_0, obtained after applying a correction for the ionic strength, I. The acid strength increases with temperature. Nitrous acid is evidently much weaker than nitric acid, for which $pK_0 = -1.44$ at $25°C$. Previously reported values of pK are greater by 0.2 to 0.4, probably because of partial decomposition of the acid during measurements.

TABLE 1

The Dissociation Constant of Nitrous Acid

Temperature (°C)	I	pK	pK_0
15	0·2504	3·018	3·230
20	0·2501	2·989	3·203
25	0·2498	2·951	3·148
30	0·2491	2·890	3·113

Isomers of Nitrous Acid

The infrared spectra of gaseous mixtures containing nitrous acid show an unusual duplication of the more intense absorption bands.[11] Thus in the frequency region where one would expect one band due to the O–H stretching vibration, two intense bands are found in the spectrum of HONO. The same feature is observed in the spectrum of DONO. A similar duplication of infrared bands is observed in the spectra of alkyl nitrites which are known to exist as geometrical isomers. It has therefore been concluded that isomeric forms of nitrous acid can exist. They are believed to arise from a restriction of rotation about the N–O(H) bond and are designated Z(cis) and E(trans) respectively.

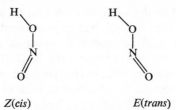

Z(cis) E(trans)

The O–H and O–D stretching frequencies (in cm^{-1}) assigned in the two isomers are:

	O–H	O–D
E-HONO	3590	2650
Z-HONO	3426	2530

Probable values for the structural parameters of both isomers have been calculated from rotational spacings observed in the infrared spectrum. More accurate values for E-HONO have been obtained by microwave spectroscopy.[12] These data are given in Table II. Both isomers appear to be planar. From measurements of the Stark effect on

TABLE II

Structural Parameters of Isomers of Nitrous Acid

	Bond lengths (nm)			Bond angles (deg.)	
	O–H	N–O	N–O	N–O–H	O–N–O
Z-HONO (i.r.)	0·098	0·146	0·120	103°	114°
E-HONO (i.r.)	0·098	0·146	0·120	105°	118°
(microwave)	0·0954	0·1433	0·1177	102·3°	110°39′

the microwave spectrum, the dipole moment of E-HONO has been calculated as 1·86 ± 0·06 D, directed at 41°24′ to the a-axis (Fig. 6.1).

The O,H bond length and N–O–H angle of E-HONO are very close to the values of corresponding parameters in nitric acid (0·0964 nm and 102°9′ respectively) and in form-aldioxime (0·0956 nm and 102°41′). The N,O bond length of 0·1433 nm is slightly less than that expected for a N,O single bond and the terminal N,O bond length of 0·1177 nm is a little shorter than the value usually accepted for a N,O double bond (0·119 nm).

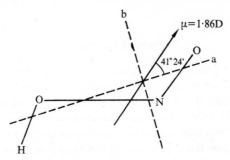

FIG. 6.1. Structure of E-HONO determined by microwave spectroscopy.

The six fundamental vibrational modes of each nitrous acid isomer are, according to the C_s point group to which they belong, five in-plane modes (ν_1 to ν_5) and one out-of-plane torsion mode, ν_6. The in-plane modes, in order of decreasing frequency, can be described as the O–H stretch, the N=O stretch, the N–O–H angle deformation, the N–O stretch and the O–N–O angle deformation. In the spectrum of nitrous acid vapour containing both isomers, two bands should be observed in each of the frequency regions where these fundamentals occur. All such bands have been seen and identified[13] and, on the basis of the dependence with temperature of their intensities, assigned to either the Z or E isomer (Table III). E-HONO is believed to be stabilized relative to Z-HONO because of repulsion effects between non-bonded electrons. Assuming that the nitrogen and oxygen atoms in the planar molecule are sp^2 hybridized, electron pair repulsions will be greates in Z- than in E-HONO (Fig. 6.2). The energy barrier to internal rotation which separater

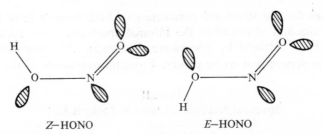

Z–HONO   E–HONO

FIG. 6.2. Bonds and non-bonding electron-pairs in E- and Z-HONO.

the two isomers has been calculated[13] as 48·4 ± 2·2 kJ mol⁻¹. This is an unusually high value in view of the essentially single-bond nature of the bond about which rotation occurs.

The existence of distinct isomers corresponding to different angles of rotation has been recognized for many years in organic chemistry. For example, internal rotation about the C–C bond in ethane is hindered and substitution at both ends of the molecule in an un-

TABLE III
Infrared Absorption Bands of E and Z Nitrous Acid observed[15] in Gaseous Mixtures of NO, NO_2 and H_2O (frequencies in cm^{-1})

	E-HONO	Z-HONO
ν_1	3605 (R) 3588 (Q) 3575 (P)	3442 (R) 3408 (P)
ν_2	1715 (R) 1699 (Q) 1684 (P)	1640 (Q)
ν_3	1278 (R) 1265 (Q) 1252 (P)	1261 (Q)
ν_4	806 (R) 791 (Q) 777 (P)	868 (R) 853 (Q) 842 (P)
ν_5	607 (R) 593 (Q) 580 (P)	608 (Q)
ν_6	540 (Q)	638 (Q)

symmetrical way can give two or more distinct equilibrium configurations. Microwave spectroscopy is an eminently suitable technique for investigating rotational isomers because the very small line widths characteristic of this region should ensure that the spectra of the different isomers are easily resolvable. Although nitrous acid has been studied[14] by this technique, the two isomers have not yet been identified in the spectrum.

The study of the interconversion of two isomers for which the energy of activation is likely to be small is facilitated by the isolation of the reactive molecules in a matrix of inert solid material at a very low temperature. Pimentel[15] has examined the $Z \rightarrow E$ conversion of nitrous acid in a solid nitrogen matrix at $-253\,°C$ by following changes in the infrared spectrum.

The photolysis of hydrazoic acid[16] in a solid nitrogen matrix at $-253\,°C$ in the presence of oxygen yields both isomers of HONO. Ultraviolet radiation causes the $E \rightarrow Z$ conversion and infrared radiation reverses this. Although chemical effects of radiation in the visible and ultraviolet regions have been recognized for many years, the isomerization of nitrous acid appears to be the first instance of a reaction induced by infrared radiation. It can be effected by irradiation within certain rather narrow frequency ranges. Both HONO and DONO isomerize[17] when matrices of these in nitrogen at $-253\,°C$ are irradiated between 3600 and $3200\,cm^{-1}$. In this range lie the O–H stretching frequencies for Z- and E-HONO (at 3410 and $3558\,cm^{-1}$ respectively in this matrix at $-253\,°C$). Careful investigation has shown that both reactions $Z \rightleftharpoons E$ can be brought about.

The mechanism of isomerization is believed to involve absorption of energy within the specified frequency range by a fundamental or combination mode to give an excited vibrational state. Intramolecular transfer of energy can occur from this to an excited torsional mode. There is a loss of energy and the molecule returns to its ground state and there is an approximately even chance that isomerization will accompany this process or that the molecule will return to its original configuration. The height of the potential barrier between the isomers has been estimated as $40 \cdot 6 \pm 2 \cdot 9$ kJ mol^{-1}.

The diffuse absorption bands in the electronic spectrum of gaseous HONO and DONO between 300 and 400 nm show contributions from both rotational isomers to the vibrational fine structure.[18] It has been estimated[11] that Z- is higher than E-HONO by 2117 J mol^{-1}. Z-HONO is destabilized by repulsion between electron pairs on adjacent atoms although the extent of destabilization is reduced by hydrogen-bond formation between the hydrogen and the terminal oxygen atom. The energy separation between Z- and E-HONO increases to 5440 J mol^{-1} in the excited electronic state. In this, the intramolecular hydrogen-bond is believed to be weakened by an electronic transition of primarily $n \to \pi^*$ type (n refers to non-bonding electrons on the terminal oxygen).

The rate constant for the reaction $Z \to E$ has been calculated[19] using the transition state model. The isomerization half-life is about 5 μsec at 25°C. This is about three orders of magnitude less than half-life of formation of HONO.

References

1. SCANLEY, C. S., *J. Amer. Chem. Soc.* **85**, 3888 (1963).
2. WAYNE, L. G., and YOST, D. M., *J. Chem. Phys.* **19**, 41 (1951).
3. WALDORF, D. M., and BABB, A. L., *J. Chem. Phys.* **39**, 432 (1963).
4. PORTER, G., *J. Chem. Phys.* **19**, 1278 (1951).
5. WALDORF, D. M., and BABB, A. L., *J. Chem. Phys.* **40**, 1476 (1964).
6. BAYLISS, N. S., and WATTS, D. W., *Aust. J. Chem.* **9**, 319 (1956).
7. SINGER, K., and VAMPLEW, P. A., *J. Chem. Soc.* 3971 (1956).
8. BAYLISS, N. S., DINGLE, R., WILKIE, R., and WATTS, D. W., *Aust. J. Chem.* **16**, 933 (1963).
9. LUMME, P., and TUMMAVUORI, J., *Acta Chem. Scand.* **19**, 617 (1965).
10. LUMME, P., LAHERMO, P., and TUMMAVUORI, J., *Acta Chem. Scand.* **19**, 2175 (1965).
11. JONES, L. H., BADGER, R. M., and MOORE, G. E., *J. Chem. Phys.* **19**, 1599 (1951).
12. COX, A. P., and KUCZKOWSKI, R. L., *J. Amer. Chem. Soc.* **88**, 5071 (1966).
13. McGRAW, G. E., BERNITT, D. L., and HISATSUNE, I. C., *J. Chem. Phys.* **45**, 1392 (1966).
14. LIDE, D. R., *Trans. Amer. Cryst. Assn.* **2**, 106 (1966).
15. PIMENTEL, G. C., *J. Amer. Chem. Soc.* **80**, 62 (1958).
16. BALDESCHWIELER, J. D., and PIMENTEL, G. C., *J. Chem. Phys.* **33**, 1008 (1960).
17. HALL, R. T., and PIMENTEL, G. C., *J. Chem. Phys.* **38**, 1889 (1963).
18. KING, G. W., and MOULE, D., *Can. J. Chem.* **40**, 2057 (1962).
19. HISATSUNE, I. C., *J. Phys. Chem.* **72**, 269 (1968).

CHAPTER 7

DIFLUORODIAZINE

Introduction

Compounds between nitrogen and fluorine have aroused much interest during the last twenty years in view of their potential usefulness as rocket propellants. The nitrogen–fluorine bond is of low energy and is easily dissociated. Substances containing it may thus be regarded as a source of fluorine without some of the difficulties associated with the handling of the element itself.

Difluorodiazine itself is a special source of interest because it can exist in two isomeric forms due to the presence of a double bond between the nitrogens which restricts the rotation of one half of the molecule relative to the other. It therefore shows geometrical isomerism of the type well-known in unsaturated organic compounds but comparatively rare in inorganic chemistry.

Preparation

Four binary compounds between nitrogen and fluorine are known. The first of these to be made was nitrogen trifluoride, NF_3, prepared[1] by Ruff in 1928 by the electrolysis of molten anhydrous ammonium hydrogen fluoride in a copper cell with a copper cathode and a carbon anode. Fluorine azide, FN_3, was made[2] in 1942 by the reaction between fluorine and hydrazoic acid and this was converted to difluorodiazine, N_2F_2, by thermal decomposition. Subsequently, tetrafluorohydrazine, N_2F_4, has been synthesized by the thermal reaction of NF_3 with various fluoride acceptors like copper, bismuth, arsenic or stainless steel.

Fluorine azide, a greenish-yellow gas at ordinary temperatures, decomposes slowly at room temperature and rapidly at 100°C to give N_2F_2 and N_2. The decomposition often occurs explosively and other more convenient syntheses have been proposed.

In 1947, Bauer[3] studied the electron diffraction of small samples of gaseous N_2F_2 prepared from fluorine azide and found his results were best accounted for by the presence of two isomeric forms of N_2F_2. These were identified as geometrical isomers arising from the presence of the central double bond. They are represented in Fig. 7.1 and are commonly referred to in the literature as *cis*- and *trans*-N_2F_2. In systematic nomenclature recently recommended[4] for isomeric systems, these are respectively the Z and E forms.

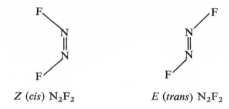

Z (*cis*) N_2F_2 E (*trans*) N_2F_2

FIG. 7.1. The isomers of difluorodiazine.

The existence of a third isomer, 1,1-difluorodiazine, in which both fluorines are substituted on the same nitrogen, has been proposed, but there has been little experimental evidence to support this and recent structural investigations have established beyond reasonable doubt that N_2F_2 exists solely in the Z and E forms.

N_2F_2 is formed[5] as a by-product (between 5 and 10% of the gases condensable by liquid nitrogen) in the electrolysis of molten ammonium hydrogen fluoride. A mixture of the isomers is separated from NF_3 and other gaseous products by fractional distillation. E-N_2F_2 is obtained in high purity from the isomeric mixture by chromatography on a molecular sieve column. Pure Z-N_2F_2 has been isolated by low-temperature distillation.

Pure E-N_2F_2 has been made[6] in about 45% yield from N_2F_4 and $AlCl_3$:

$$3N_2F_4 + 2AlCl_3 \rightarrow 3N_2F_2 + 3Cl_2 + 2AlF_3$$

Some elementary nitrogen is also formed according to:

$$3N_2F_4 + 4AlCl_3 \rightarrow 3N_2 + 6Cl_2 + 4AlF_3$$

Optimum yields of N_2F_2 are achieved by reaction for short periods (0·5 to 5 min) between gaseous N_2F_4 and solid $AlCl_3$ at $-80°C$. Unchanged N_2F_4 in the product is destroyed by shaking it with water in the presence of oxygen and a trace of nitric oxide.

A convenient laboratory synthesis[7] uses urea and fluorine as starting materials. Fluorine, diluted with nitrogen to moderate the reaction, combines with urea at $0°C$ to give a yellow, corrosive liquid which contains difluoroamine, HNF_2. This may be isolated by distillation. The second stage is the reaction between HNF_2 and an alkali metal fluoride, MF (where M = K, Rb or Cs). Addition compounds of formula, $MF \cdot HNF_2$, are first formed but these decompose on heating to give N_2F_2. The overall reaction is

$$2MF + 2HNF_2 \rightarrow 2MF \cdot HF + N_2F_2$$

N_2F_2 has also been made[8] by the action of aqueous base on difluoro-urea. At temperatures below $25°C$ and using a concentrated solution of an alkali metal hydroxide, yields of N_2F_2 between 20 and 40% are obtained. The reaction of fluorine with sodium azide has been studied[9] by Russian workers. The product is a mixture of fluorides of nitrogen, nitrous oxide and nitrogen. Chromatographic analysis using silica gel gave the typical composition: Z-N_2F_2, 21·2%; E-N_2F_2, 47·0%; NF_3, 18·7%. N_2F_2 is formed via the intermediate FN_3.

$$NaN_3 + F_2 \rightarrow FN_3 + NaF$$
$$FN_3 \rightarrow \tfrac{1}{2}N_2F_2 + N_2$$

E-N_2F_2 melts at $-172°$ and boils at $-111·4°C$. Z-N_2F_2 melts below $-195°$ andboils at $-105·7°C$.

Reactions

Z-N_2F_2 is a much more reactive compound than E-N_2F_2. Thus it reacts slowly with glass at room temperature to form SiF_4 and N_2O, whereas E-N_2F_2 shows no reaction after being sealed in a glass tube at room temperature for a month.

Difluorodiazine exerts a catalytic effect on the polymerization of methyl methacrylate, styrene and cyclopentadiene. Under a pressure of $40 \, kNm^{-2}$ of N_2F_2, each of these monomers is polymerized at room temperature within twelve hours. Z-N_2F_2 is a more effective catalyst than E-N_2F_2.

In its reactions, Z-N_2F_2 behaves[10] either as a simple fluorinating agent or as a combined deoxygenating and fluorinating agent. With SO_2, reaction at $100°C$ gives a mixture of SO_2F_2, SOF_2, N_2O and N_2. With POF_3, the products are PF_5, N_2 and O_2. SF_4 and PF_3 are fluorinated to give SF_6 and PF_5 respectively.

Both isomers react with SbF_5 to form a solid of composition, $NF \cdot SbF_5$.[11] On the basis of its infrared spectrum, this compound is formulated as an ionic substance $(N_2F^+) \cdot (Sb_2F_{11})^-$. The solid melts between $82°$ and $84°$ and decomposes at $200°C$ to N_2F_2. Not all the N_2F_2 originally taken can be recovered. Essentially pure Z-N_2F_2 is obtained regardless of which isomer was used in the preparation. Equimolar amounts of Z-N_2F_2 and AsF_5 react at $-196°C$ to give a similar $1:1$ adduct, formulated as $(N_2F^+)(AsF_6^-)$.[12] This is a white solid, stable at room temperature. E-N_2F_2 does not react with AsF_5 at low temperatures.

Isomerization

The heat of isomerization has been determined from calorimetric measurements.[13] The heat of formation of each isomer was measured by ignition with excess ammonia in a bomb calorimeter. The products of this reaction were ammonium fluoride and nitrogen and hydrogen gases. The heat of isomerization $Z \rightarrow E$ was found to be $12\cdot55 \pm 1\cdot25 \, kJ$ mol^{-1}, the Z isomer being the more stable of the two.

Another investigation of the interconversion of the two isomers[14] showed that the equilibrium constant

$$K = \frac{[Z-N_2F_2]}{[E-N_2F_2]}$$

was independent of temperature over the range $25°$ to $150°C$, its average value being $9\cdot5$. According to these results, the heat of isomerization must be zero. In this case, samples were kept in stainless steel ampoules and it is very probable that the isomerization was catalysed by the walls of these vessels.

The process has also been studied under homogeneous conditions. To achieve these, E-N_2F_2, highly diluted with argon, was heated by reflected shock waves to elevated temperatures, at which it was maintained and allowed to react for a short period of time (~ 2 msec) and then rapidly cooled by an expansion wave. The tube walls remained cold and so heterogeneous catalysis could not occur. Using this technique, the conversion E- to Z-N_2F_2 was studied[15] between $297°$ and $342°C$ and at pressures up to $160 \, kNm^{-2}$. The equilibrium constant was found to be approximately 20 over this temperature range and the activation energy for isomerization calculated as approximately $350 \, kJ \, mol^{-1}$. This is the energy required to decouple, by rotation, the π-bond overlap between the two nitrogens. It is significantly greater than the activation energy, about 250 kJ mol⁻¹, required to bring about comparable rotation in C=C bonds.

Molecular Structure

Bond lengths and bond angles determined for E- and Z-N_2F_2 by electron diffraction[17] together with values for Z-N_2F_2 found by microwave spectroscopy[16] are given in Table I. For Z-N_2F_2, the values of the N,N bond length and the FNN bond angle determined by the two methods are the same, but values for the N,F bond length do not agree. In both

TABLE I

Molecular Dimensions of Difluorodiazine Isomers

Isomer	N,N nm	N,F nm	FNN angle
E-N_2F_2*	0·123	0·1396	105°30′
Z-N_2F_2*	0·1214	0·141	114°24′
Z-N_2F_2†	0·1214	0·1384	114°28′

* Electron diffraction results.[17]
† Microwave spectra results.[16]

isomers, the N,N bond length is found to be shorter than in comparable molecules with N,N double bonds (cf. 0·124 nm in $CH_3N=NCH_3$ and 0·125 nm in $C_6H_5N=NC_6H_5$). The major structural difference between the two isomers is the greater FNN bond angle in Z- compared with E-N_2F_2. This is understandable in view of the juxtaposition of the two fluorines in Z-N_2F_2 which leads to increased repulsion between them. The discrepancy in the values of the N,F bond length for Z-N_2F_2 precludes any valid comparison with the corresponding parameter in E-N_2F_2. However, the difference between the N,N bond length in the two molecules appears to be quite real and we must conclude that isomerization in N_2F_2 is accompanied by a significant change in the length of this bond.

The dipole moment of Z-N_2F_2, determined from the microwave spectrum, is 0·16 D. This is small in view of the large polarity expected for a N,F bond. It does, however, parallel the small moments of NF_3 and N_2F_4 (0·24 and 0·27 D respectively) and can be attributed, as in these compounds, to counteraction of the polarity of the N–F bond by the hybridized lone-pair on the nitrogen.

In view of its symmetry, E-N_2F_2 is a non-polar molecule.

Vibrational Spectra

For Z-N_2F_2, belonging to the point group C_{2v}, there are six normal modes of vibration: three of A_1 and two of B_2 symmetry (all five allowed in both the infrared and Raman spectra) and a torsional mode of species A_2 which is Raman-active but infrared forbidden (Fig. 7.2). An assignment of these has been made[18] using Raman data on the liquid and infrared data on the gas. In Table II, this is compared with an assignment for E-N_2F_2.[19] The band at 550 cm⁻¹ in the Raman spectrum of Z-N_2F_2 is very weak and its identification with the infrared-inactive torsional mode, A_2, is tentative, particularly as a band is observed near 550 cm⁻¹ in the infrared spectrum. It may be that some as yet unidentified impurity

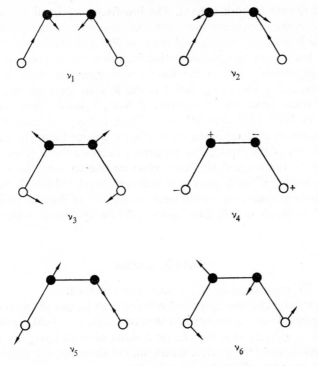

FIG. 7.2. The six normal modes of vibration of Z-N_2F_2 (in the point group C_{2v}).

TABLE II

Vibrational Assignments for Z- and E-N_2F_2

Z-N_2F_2

Class	Description	Vibration	Frequency (cm^{-1})
A_1	N=N, stretch	ν_1	1525
	N—F, sym. stretch	ν_2	896
	N—F, sym. bend	ν_3	341
A_2	Torsion	ν_4	550
B_2	N—F, asym. stretch	ν_5	952
	N—F, asym. bend	ν_6	737

E-N_2F_2

Class	Description	Vibration	Frequency (cm^{-1})
A_g	N=N, stretch	ν_1	1522
	N—F, sym. stretch	ν_2	1010
	N—F, sym. bend	ν_3	600
A_u	Torsion	ν_4	363·5
B_u	N—F, asym. stretch	ν_5	990
	N—F, asym. bend	ν_6	423

gives rise to this feature in both spectra. The low-frequency band at 341 cm^{-1} is in line with observations in the microwave spectrum[16] of rotational transitions in the ground and excited states which indicate a vibrational mode at 300 ± 35 cm^{-1}.

The E-isomer has a centrosymmetric molecule, belonging to the point group, C_{2h}. As a consequence, vibrations active in the Raman spectrum and those in the infrared are mutually exclusive. v_1, v_2 and v_3 are active in the Raman spectrum but forbidden in the infrared. The Raman spectrum of gaseous E-N$_2$F$_2$[20] shows three relatively intense, polarized bands at 1018, 1523 and 603 cm^{-1}. These correspond respectively with v_2, v_1 and v_3. The frequencies are very close to, if slightly greater than, the values observed in the Raman spectrum of the liquid. The frequency decrease for a given band on passing from the gaseous to the condensed phase could be due to some kind of molecular association, a 'fluorine-bond' analogous to a hydrogen-bond. Although the molecule as a whole has no dipole moment, the strong electronegativity of fluorine means that the N–F bond is polar and so dipole–dipole interaction could be significant in the condensed state

N.M.R. Spectra

Both ^{14}N and ^{19}F resonances in N$_2$F$_2$ have been observed.

The ^{14}N nuclear magnetic resonance spectrum of each isomer shows only a single peak. This appears to preclude an unsymmetrical structure such as 1,1-difluorodiazine for either isomer because it is unlikely that the chemical shifts of both nitrogens in this would be the same. High-resolution ^{14}N spectra, which should show fine structure due to ^{14}N/^{19}F and ^{14}N/^{14}N spin-coupling effects have not been reported yet.

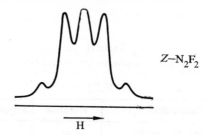

$$Z–N_2F_2$$

$$H \longrightarrow$$

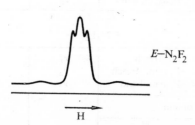

$$E–N_2F_2$$

$$H \longrightarrow$$

FIG. 7.3. ^{19}F nuclear magnetic resonance spectra of Z-N$_2$F$_2$ and E-N$_2$F$_2$ at 40 MHz (taken with permission from *J. Chem. Phys.* **37**, 182 (1962)).

The ^{19}F spectra of both isomers at 40 MHz are symmetric five-line patterns[21] (Fig. 7.3). For Z-N_2F_2, there are two strong peaks separated by 108 ± 5 Hz from a strong central peak and two weak peaks located at $\pm 238 \pm 10$ Hz from this. For E-N_2F_2, the two strong peaks are separated from the strong central peak by 63 ± 5 Hz and the two very weak peaks are at $\pm 425 \pm 25$ Hz from this.

Chemical shifts and spin coupling constants have been determined from the double resonance spectra of the two compounds. In this technique, the multiplet structure of the spectrum is removed by decoupling the spins of certain nuclei from the rest of the spin system. ^{19}F double resonance spectra are obtained by applying a strong oscillatory field, H_2, of frequency ν_2 near the resonance frequency of ^{14}N and studying the resonance of the fluorine nuclei by sweeping a weak oscillatory field, H_1, through the region where this occurs.

N_2F_2 is an $AA'XX'$ system. A and A' refer to the two ^{19}F nuclei ($I = \frac{1}{2}$) which have the same chemical shift but are magnetically non-equivalent, i.e. they have different coupling constants to the nitrogen nuclei. X and X' refer to ^{14}N nuclei ($I = 1$) which have the same chemical shift as each other, but are again magnetically non-equivalent.

When the double resonance spectra of $AA'XX'$ systems are recorded, the resonance in the A region of the spectrum is symmetric provided that the frequency ν_2 is equal to the resonance frequency, ν_X, of the X nucleus. When these are not equal, resonance in the A region is generally not symmetric, but when ν_2 is set at equal frequency intervals above and below ν_X, the two spectra obtained are mirror images. Thus ν_X is found by changing ν_2 until a symmetric double resonance spectrum is obtained.

The ^{19}F and ^{14}N chemical shifts are summarized in Table III and compared with those for NF_3 and N_2F_4. The double resonance spectra of the E-isomer are those expected for a spin system of type $AA'XX'$ with C_{2h} symmetry. The spectra of the other isomer are consistent with an $AA'XX'$ system (e.g. Z-N_2F_2 with C_{2v} symmetry) or an A_2XX' system (e.g. 1,1-difluorodiazine). However, in view of the conclusions reached from other spectroscopic studies and the electron diffraction work, it is clear that the N.M.R. spectra should be interpreted according to the first of these systems.

TABLE III

N.M.R. Spectroscopic Data for Nitrogen Fluorides. The N_2F_2 molecule is regarded as F—N=N′—F′

Compound	Chemical shift (ppm)		Coupling constant (Hz)		
	^{19}F	^{14}N	J_{NF}	$J_{N'F}$	$J_{FF'}$
E-N_2F_2	− 94·9	−75·2	±136	∓73	322
Z-N_2F_2	−133·7	−13·3	±145	∓37	99
NF_3	−146·9	0·0			
N_2F_4	− 59·8	+39·8			
CCl_3F	0				

References

1. RUFF, O., FISCHER, J., and LUFT, F., *Zeit. anorg. all. Chem.* **172**, 417 (1928).
2. HALLER, J. F., Thesis, Ph. D., Cornell University, 1942.
3. BAUER, S. H., *J. Amer. Chem. Soc.* **69**, 3104 (1947).
4. BLACKWOOD, J. E., GLADYS, C. L., LOENING, K. L., PETRARCA, A. E., and RUSH, J. E., *J. Amer. Chem. Soc.* **90**, 509 (1968).

5. COLBURN, C. B., JOHNSON, F. A., KENNEDY, A., McCALLUM, K., METZGER, L. C., and PARKER, C. O., *J. Amer. Chem. Soc.* **81**, 6397 (1959).
6. HURST, G. L., and KHAYAT, S. I., *J. Amer. Chem. Soc.* **87**, 1620 (1965).
7. LAWTON, E. A., PILIPOVICH, D., and WILSON, R. D., *Inorg. Chem.* **4**, 118 (1965).
8. JOHNSON, F. A., *Inorg. Chem.* **5**, 149 (1966).
9. PANKRATOV, A. V., SOKOLOV, O. M., and SAVENKOVA, N. I., *Zhur. Neorg. Khim.* **8**, 2030 (1964).
10. LUSTIG, M., *Inorg. Chem.* **4**, 104 (1965).
11. RUFF, J. K., *Inorg. Chem.* **5**, 1791 (1966).
12. MOY, D., and YOUNG, A. R., *J. Amer. Chem. Soc.* **87**, 1889 (1965).
13. ARMSTRONG, G. T., and MARANTZ, S., *J. Chem. Phys.* **38**, 169 (1963).
14. PANKRATOV, A. V., and SOKOLOV, O. M., *Zhur. Neorg. Khim.* **11**, 1761 (1966).
15. BINENBOYM, J., BURCAT, A., LIFSHITZ, A., and SHAMIR, J., *J. Amer. Chem. Soc.* **88**, 5039 (1966).
16. KUCZKOWSKI, R. L., and WILSON, E. B., Jr., *J. Chem. Phys.* **39**, 1030 (1963).
17. BOHN, R. K., and BAUER, S. H., *Inorg. Chem.* **6**, 309 (1967).
18. KING, S-T., and OVEREND, J., *Spectrochim. Acta* **23** A, 61 (1967).
19. KING, S-T., and OVEREND, *Spectrochim. Acta* **22** A, 689 (1966).
20. SHAMIR, J., and HYMAN, H. H., *Spectrochim. Acta* **23** A, 1191 (1967).
21. NOGGLE, J. H., BALDESCHWIELER, J. D., and COLBURN, C. B., *J. Chem. Phys.* **37**, 182 (1962).

CHAPTER 8

PHOSPHORUS PENTACHLORIDE

Introduction

Representative elements which follow nitrogen in Group V form two distinct series of compounds in which the element shows oxidation states respectively of +3 and +5. This is exemplified in the formation of tri- and pentahalides with a particular halogen.

Studies of the halides of Group V elements have played an important part in the development of valence theory. The formation of a pentahalide can be accounted for by the involvement in bonding of a *d*-orbital as well as the *s*- and *p*-orbitals of the valence shell. Such expansion of the valence shell is a necessary prerequisite for the formation of compounds by Group V, VI and VII elements in which the element shows the same valence as the number of its Periodic Group.

A feature of additional interest shown by some pentahalides is their ability to assume different structural forms, depending on their physical state. Phosphorus pentachloride is the most thoroughly investigated compound of this kind.

Preparation

Like the halogen compounds of Group V elements in general, phosphorus pentachloride is sensitive to water and must be prepared and studied in anhydrous conditions. It can be made by the interaction of excess chlorine with phosphorus or, under more easily controlled conditions, by the direct combination of phosphorus trichloride and chlorine.[1] When phosphorus trichloride liquid is dropped into a reaction vessel filled with chlorine gas, the pentachloride is formed immediately.

The solid is a very pale yellow colour: it melts under pressure at 148°C and sublimes at 160°C under atmospheric pressure. The vapour contains PCl_5 molecules which can dissociate according to:

$$PCl_5(g) \rightleftharpoons PCl_3(g) + Cl_2(g)$$

At 160°C, the vapour is 13·5% dissociated. This system has been extensively studied as an example of chemical equilibrium in the gas phase.

Reactions

Phosphorus pentachloride is a very reactive compound. It behaves as a chlorinating agent due to the ease with which chlorine is formed by dissociation of the molecule. Thus it reacts with many metals to form the corresponding metal chloride. The reaction between phosphorus pentoxide and phosphorus pentachloride produces phosphorus oxochloride, $POCl_3$. PCl_5 rapidly absorbs moisture to form a mixture of HCl and H_3PO_4:

$$PCl_5 + 4H_2O \rightarrow 5HCl + H_3PO_4$$

With a number of metal fluorides, PCl_5 reacts to give the metal salt of hexafluorophosphoric acid. For example, reaction with KF gives an 80% yield of KPF_6:

$$PCl_5 + 6KF \rightarrow KPF_6 + 5KCl$$

It is used extensively in organic chemistry as a chlorinating agent. In reaction with olefins, the first product from $RCH=CH_2$ is the adduct $RCHCl-CH_2PCl_4$. This can be converted to an unsaturated phosphonic acid, $RCH=CHPO(OH)_2$, by treatment with water or alkali.

PCl_5 also forms addition compounds with the halides of various other elements. Typical examples are $PCl_5 \cdot BCl_3$, $PCl_5 \cdot AlCl_3$ and $PCl_5 \cdot FeCl_3$. $PCl_5 \cdot AlCl_3$ is a white powder which is soluble in nitrobenzene and acetonitrile but not in non-polar solvents like benzene. Its solution in nitrobenzene is a good conductor of electricity and the behaviour on electrolysis indicates phosphorus is in the cation and aluminium is in the anion. The adduct is accordingly formulated as $[PCl_4]^+[AlCl_4]^-$. Probably this ionic type of structure is present in other addition compounds of PCl_5.

Structure in the Vapour State

Electron diffraction studies of the vapour of PCl_5[2] and similar compounds, including PF_5, PF_3Cl_2, AsF_5 and $SbCl_5$, have shown that all these molecules have a trigonal bipyramidal shape. In PCl_5, the bonds lengths are not all equal. The internuclear distance between phosphorus and the chlorine atoms in the equatorial plane of the bipyramid is 0.204 ± 0.006 nm. The corresponding distance relating to the axial chlorines, lying on the three-fold molecular axis above and below the equatorial plane, is 0.219 ± 0.002 nm.

Structure in the Liquid State

The vibrational spectrum of any AX_5 molecule should provide a clear differentiation between the two probable configurations, either a trigonal bipyramid (D_{3h} symmetry) or a tetragonal pyramid, C_{4v}. If it is a trigonal pyramid, the eight normal vibrations are grouped into the species $2A_1' + 2A_2'' + 3E' + E''$. Six of these ($2A' + 3E' + E''$) are Raman-active and five ($2A_2'' + 3E'$) are infrared active. If it is a tetragonal pyramid, all nine normal vibrations are Raman active but only six are infrared active. It is therefore possible to distinguish between these two configurations in the Raman and infrared spectra.

The normal modes of vibration of a trigonal bipyramidal AX_5 molecule[3] are illustrated in Fig. 8.1.

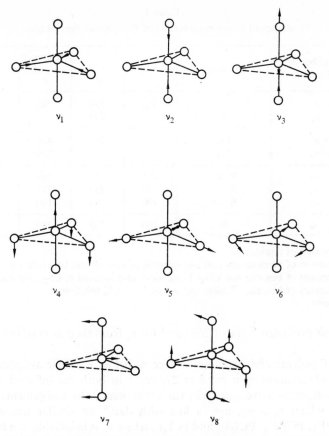

FIG. 8.1. Normal vibrational modes of a trigonal bipyramidal molecule like
PCl$_5$.

The first measurements of the Raman spectrum[4] of liquid phosphorus pentachloride showed six lines, believed to be those predicted for D_{3h} symmetry. They were located at 100, 190, 271, 392, 449 and 495 cm^{-1}. Not all these lines were found in subsequent investigations and it now appears that some may have been due to impurities such as phosphorus oxochloride or trichloride because insufficiently rigorous precautions were taken to purify the PCl$_5$ sample or to prevent access of moisture.

Table I summarizes the fundamental vibrational frequencies assigned by various investigators.[3,5,6,7,8] Some of the first assignments are not reliable. For example, the Raman line at 495 cm^{-1} observed by Siebert is believed to be due to phosphorus oxochloride. There is good agreement between most of the assignments made in recent studies.

The main discrepancies concern ν_2 and the relative positions of ν_6 and ν_8. Carlson[6] observed only one of the Raman-active, polarized A_1 vibrations (that at 393 cm^{-1}) and concluded that the second of these was coincident with the first. A coincidence is not likely because it would mean the stretching force constants of the axial and equatorial bonds were of the same magnitude. This would be contrary to the electron diffraction results which showed quite clearly the non-equivalence of these bonds. Taylor and Woodward[7] identified a weak Raman line at 370 cm^{-1} with ν_2 but Condrate and Nakamoto[8] have assigned the band they observed at 264 cm^{-1} to this vibration because this frequency is

73

TABLE I

Fundamental Frequencies (cm^{-1}) of Phosphorus Pentachloride

Class	Number of vibration	Activity	Reference				
			3	5	6	7	8
A_1'	ν_1	R	392	394	393	395	394
	ν_2	R	271	394		370	264
A_2''	ν_3	i.r.		465	446	441	465
	ν_4	i.r.			299	301	299
E'	ν_5	R, i.r.	495	592	580	581	592
	ν_6	R, i.r.	190	335	273	281	273
	ν_7	R, i.r.	100	100	100	100	100
E''	ν_8	R	190	280	280	261	282

Ref. 3. Raman spectrum of liquid.
Ref. 5. Raman spectrum of benzene solution and infrared of vapour and CS_2 solution.
Ref. 6. Raman spectrum of benzene and CH_2Cl_2 solutions and infrared of CS_2 solution.
Ref. 7. Raman spectrum of benzene solution and infrared of CCl_4 solution.
Ref. 8. Reassignments.

very close to their predicted value of 260 cm^{-1} for ν_2 from their normal coordinate analysis of PCl_5.

Carlson and Condrate and Nakamoto place ν_6 below ν_8, their assignment of ν_6 being based on their observation of a band at 273 cm^{-1} in both the infrared and the Raman spectrum. ν_8 is Raman-active only. On the other hand, the assignments of Taylor and Woodward, in which $\nu_6 > \nu_8$, are in line with data[9] on similar trigonal bipyramidal molecules like PF_5, PCl_2F_3, PCl_3F_2 and PCl_4F, where ν_6 is invariably at a higher frequency than ν_8.

Although there are these differences in detailed assignments of the fundamental vibrations of PCl_5, there is no doubt that the vibrational spectrum is completely in accord with a trigonal bipyramidal structure for the molecule in the liquid and in solution in non-polar solvents.

Structure of the Solid

The molecular nature of PCl_5 in the vapour and liquid states is clearly established in the work so far described but it is very difficult to reconcile this with the physical properties of the solid. These are consistent with much stronger forces between the structural units than are usual between covalent molecules.

The first indication of the kind of structural units present in the solid came from the observation[10] that there is a marked change in Raman spectrum on going from the liquid to the solid phase. Four lines (at frequencies 248, 357, 405 and 450 cm^{-1}) were found for the solid and these were attributed to the presence of PCl_4^+ tetrahedra, suggesting the structure must be ionic. Support for this comes from the electrical conductivity shown by the solid in contrast to the behaviour of the liquid as a non-conductor.

The X-ray crystal analysis of PCl_5 was reported in 1942.[11] To prevent the rapid decomposition of phosphorus pentachloride when exposed to an atmosphere containing water vapour, single crystals for X-ray analysis were grown from nitrobenzene solution and then coated with a paraffin and vaseline mixture to preserve them during examination. The solid contains two kinds of phosphorus atom, crystallographically different from each other. It is composed of PCl_4^+ tetrahedra and PCl_6^- octahedra arranged in a lattice which is essentially of the caesium chloride type but with some distortion due to the non-spherical shape of the ions. The average P,Cl distance in PCl_6^- is 0·206 nm, that in PCl_4^+ is 0·198 nm.

Solutions of PCl_5 in polar solvents such as $POCl_3$, $C_6H_5NO_2$, C_6H_5COCl and CH_3COCN show[12] electrical conductivity because the ionization

$$2PCl_5 \rightleftharpoons PCl_4^+ + PCl_6^-$$

is promoted. The extent of ionization appears to be determined more by the ability of the solvent to stabilize the ions by complex formation than by its dielectric constant. These ions have been recognized individually in other compounds such as $PCl_5 \cdot AlCl_3$, which has the ionic structure $[PCl_4]^+[AlCl_4]^-$ and $AsCl_5 \cdot PCl_5$ which contains $AsCl_4^+$ and PCl_6^- ions. It is apparent that the phosphorus pentachloride molecule can act as a chloride ion donor or acceptor.

A re-investigation[13] of the Raman spectrum of solid PCl_5 showed some additional lines beside those due to PCl_4^+. The extra ones must be associated with vibrations of PCl_6^-, which has six fundamental vibrations, three being active in the Raman spectrum.

The molecular form of phosphorus pentachloride has been 'frozen' into a solid phase by deposition from the vapour[6] on to a cold plate at $\sim -183°C$. The solid contains PCl_5 molecules, which may be identified by their characteristic infrared spectrum. On allowing the sample to warm up to room temperature, marked changes in the spectrum are observed (Fig. 8.2).

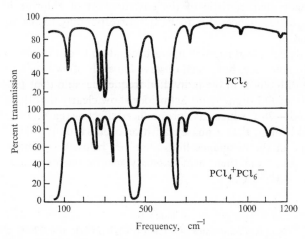

Fig. 8.2. Infrared spectrum of solid PCl_5 in the covalent and ionic forms (reproduced from Carlson, G. L., *Spectrochimica Acta* **19**, 1291 (1963)).

The bands due to PCl_5 molecules disappear and the spectrum becomes that of $[PCl_4]^+ \cdot [PCl_6]^-$. The Raman and infrared spectra of 'ionic' PCl_5 have been compared with the spectra of $[PCl_4]^+[AlCl_4]^-$ to facilitate assignment of fundamental frequencies to the PCl_4^+ and PCl_6^- ions. Those due to PCl_4^+ are at 171 (R and i.r.), 459 (R) and 658 (R and i.r.) cm^{-1}. The Raman-active frequencies due to PCl_6^- are at 150, 281 and 360 cm^{-1}.

The ^{31}P nuclear magnetic resonance spectrum[14] of solid PCl_5 confirms the presence of phosphorus atoms in two different environments.[14,15] Two resonance signals are observed in the PCl_5 spectrum studied at 26 MHz. Chemical shifts relative to 85% phosphoric acid[16] are -91 ppm for PCl_4^+ and $+282$ ppm for PCl_6^-. For the highly symmetrical PCl_6^- ion, the paramagnetic contribution to the chemical shift should be small, resulting in a strong positive shift. Measurements of the ^{31}P resonance of solutions of PCl_3 and PCl_5 in carbon disulphide have shown there is the following monotonic sequence of chemical shifts (in ppm): PCl_3, -215; PCl_4^+, -91; PCl_5, $+80$; and PCl_6^-, $+282$.

The solid has also been examined by the technique of nuclear quadrupole resonance spectroscopy.[17,18] ^{35}Cl and ^{37}Cl nuclei both have a spin of 3/2 and, in an axially symmetric field, a single transition is expected with a nuclear quadrupole resonance frequency,

$$v = \left(\frac{eQq}{2h}\right)\sqrt{1+\eta^2/3}$$

where q is the field gradient at the nucleus due to the electron distribution in the molecule or ion and eQq is the nuclear quadrupole coupling constant. q has components q_{xx}, q_{yy} and q_{zz} along the x, y and z axes respectively and η, the asymmetry parameter, $= \dfrac{q_{xx} - q_{yy}}{q_{zz}}$. For an axially symmetric field gradient, $q_{xx} = q_{yy} \neq q_{zz}$ and $\eta = 0$.

X-ray structural analysis has demonstrated that at room temperature the solid is composed of PCl_4^+ and PCl_6^- ions. There should be field gradient tensors for each of the four chlorine atoms in the cation which differ only in their spatial orientations but not in their principal values. For the anion, there should be two different field gradient tensors, one for the axial and one for the equatorial chlorines, due to its tetragonal distortion.

The predicted N.Q.R. spectrum of ionic PCl_5 should show a single resonance due to PCl_4^+ at a frequency characteristic of the environment of chlorine in it and, at lower frequencies, a pair of resonances due to PCl_6^- corresponding to the equatorial and axial chlorines in this ion.

No resonance is detectable at room temperature and no lines are observed above $-73°C$. At $-196°C$, six resonance lines have been observed in each of two independent studies,[17,18] but there are some differences in the individual frequencies and their proposed assignments (Table II). The data of DiLorenzo and Schneider show clearly that there is a phase change between $-196°C$ and $-78°C$ which has a pronounced effect on the N.Q.R. spectrum. Chihara, Nakamura and Seki found that a phase transition occurs around $-163°C$, indicated by the coalescence of some of the resonance lines observed below this temperature. Evidence in support of this comes from their heat capacity measurements which indicates a small λ point anomaly at $-170·8°C$.

TABLE II

N.Q.R. Frequencies of ^{35}Cl in $[PCl_4]^+[PCl_6]^-$ at $-196°C$

DiLorenzo and Schneider[17]			Chihara et al.[18]		
Frequency (MHz)	Resonance line	Species	Frequency (MHz)	Resonance line	Species
29·61	v_1	PCl_6^-	28·424	v_1	PCl_6^-
30·07	v_2	PCl_6^-	29·720	v_2	PCl_6^-
30·45	v_3	PCl_6^-	30·478	v_3	PCl_6^-
30·62	v_4	PCl_6^-	32·288	v_4	PCl_4^+
32·28	v_5	PCl_4^+	32·396	v_5	PCl_4^+
32·61	v_6	PCl_4^+	32·620	v_6	PCl_4^+

The N.Q.R. spectrum of solid PCl_5 in the molecular form (obtained by condensation at $-196°C$ from the vapour) shows only three resonance lines. One, at $33·75$ MHz, has been assigned to the equatorial chlorines and the other two, at about $29·2$ MHz, to the axial chlorines. The spectrum of the molecular solid becomes less intense as the temperature is raised over the range $-163°$ to $-83°C$. It is evidently metastable with respect to the ionic solid, the formation of which is readily followed by the development of the characteristic N.Q.R. spectrum. Differential thermal analysis shows there is an exothermic reaction involving the molecular solid between $-108°$ and $-59°C$. This is most probably the disproportionation:

$$2PCl_5 \rightarrow [PCl_4]^+[PCl_6]^-$$

Further investigations on solid PCl_5 such as X-ray diffraction at low temperatures and more extensive heat capacity measurements are necessary to achieve a more complete understanding of the different phases which can exist and the transitions between them.

Bonding

Many molecules of general formula AX_5 have a trigonal bipyramidal shape. On theoretical grounds,[19] the more probable arrangement of five electrons with parallel spins in five atomic orbitals of A, is trigonal bipyramidal rather than square-based pyramidal. Hence the shape of the AX_5 molecule. From the viewpoint of stereochemistry described in terms of interaction between electron pairs, repulsion between five bond pairs of electrons is minimal when they are in the trigonal bipyramidal arrangement.[20]

According to the valence bond theory of directed orbitals, the bonding in AX_5 is described in terms of a set of sp^3d hybrid orbitals. In view of the high energy ($1·3-1·6$ aJ) of the $3d$- relative to the $3s$- and $3p$-orbitals of phosphorus, it is likely that these are the preferred hybrids rather than any alternative with greater d-orbital contribution.

The power to form a bond associated with a hybrid orbital is proportional, according to Pauling, to the value of the angular part of the wave-function along the bond direction. The strength of the bond is then assumed to be proportional to the products of the bond-forming powers of the two orbitals involved. On Pauling's scale, the bond-forming power of an s-orbital is taken as unity and the bond-forming powers of the equatorial and axial bonds calculated[21] as $2·249$ and $2·937$ respectively. The inference from these figures is that equatorial bonds should be weaker, and therefore longer, than axial bonds. This is the converse of the conclusions from the electron diffraction study of PCl_5 vapour. Moreover, a normal coordinate analysis[22] of PCl_5 leads to force constant data which also indicate that the equatorial chlorines are more strongly bound than the axial. The stretching force constant is a measure of the strength of a bond at internuclear distances close to the equilibrium value. The calculated force constant for the equatorial bonds is $230·7$ Nm^{-1} and that for the axial bonds is $218·3$ Nm^{-1}.

A modified version of the valence bond description involves the use by the phosphorus atom of two different sets of hybrid orbitals. Those which form bonds in the equatorial plane are constructed from the $3s$- and two of the $3p$-orbitals and those which form the axial bonds from a combination of $3p$- and $3d$-orbitals. The two sets have different energies because of the energy required for the promotion of an s-electron to a d-level and hence different bond-forming powers. This version is rather arbitrary and difficult to justify theoretically although it does lead to predictions in agreement with experimental results.

Another criterion for bond strength is the magnitude of the overlap integral. For two orbitals ϕ_1 and ϕ_2, this is given by $S_{12} = \int \phi_1\phi_2 dv$, where the integration is made over all space. Bond strength is assumed to be proportional to the values of overlap integrals computed using Slater-type orbitals. It has been shown[23] that overlap between $sp^3d_{z^2}$-orbitals on a phosphorus atom and s- or p-orbitals on five chlorine atoms situated at the same distance from the phosphorus is greater for an equatorial than for an axial bond. The former should therefore be stronger than the latter. The argument cannot be carried a stage further to conclude that the axial should be longer than the equatorial bonds because this would directly contradict the assumption initially made that all bond lengths are equal.

Coulson[24] has discussed the extent to which d-electrons participate in bonding within molecules like PCl_5 and SF_6. Effective hybridization occurs between atomic orbitals having similar size. In the isolated phosphorus atom, the average values of radial distances in the configuration $3s3p^33d$ have been calculated[25] as $r_{3s} = 0.047$ nm, $r_{3p} = 0.055$ nm, and $r_{3d} = 0.243$ nm. It thus appears that the 3d-orbitals of phosphorus are much too large to take part in hybridization.

There are three ways[26] in which the size of a d-orbital can be altered: (i) by spin-coupling between unpaired electrons in different orbitals; (ii) by excitation of one or more electrons to d-orbitals which are unfilled in the ground state; (iii) by removal of electrons from a previously-neutral atom, leaving it with a formal positive charge. This occurs whenever the central atom is bonded to more electronegative elements. In PCl_5, the formal charge effect is particularly important because of the electronegativity of chlorine. The removal of negative charge provides a mechanism by which an electron previously excited to a 3d-orbital can contract in size and so participate in hybridization with atomic 3s- and 3p-orbitals.

Gillespie[27] has discussed the inequality of bond lengths in AX_5 molecules from the viewpoint of repulsion between electron pairs in the valence shell of the central atom. The interaction between electron pairs which make an angle of 90° with each other at the phosphorus nucleus is greater than that between electron pairs which make an angle of 120° with each other. Each equatorial pair interacts with two axial pairs only at angles of 90° whereas each axial pair interacts with three equatorial pairs. Therefore a stable arrangement with all five electron pairs at the same distance from the nucleus is impossible. The axial electron pairs should be subject to greater repulsion than the equatorial pairs and the equilibrium arrangement should have the axial pairs at a slightly greater distance than the equatorial pairs from the nucleus.

A recent theoretical treatment[28] describes the bonding in PCl_5 in terms of an orbital-deficient model, with σ-bonds formed by the s- and p-orbitals of phosphorus and only minor contributions from π-bonds involving d-orbitals. The greater strength of the equatorial over the axial bonds is related to a greater s-orbital contribution to the equatorial orbitals, leading to a higher electronegativity of these compared with the axial orbitals. There is probably some d-character in both the equatorial and axial bonds and a small contribution p_π-p_π and d_π-p_π bonds.

An estimate of the extent of π-bonding in PCl_5 has been made from nuclear magnetic resonance data[29] on the chemical shifts (in ppm) of ^{31}P in the following molecules: PF_5 ($\delta = +35$), PCl_5 (+81), $P(OC_2H_5)_5$ (+71), PBr_5 (+74) and $P(C_6H_5)_5$ (+89). On the a priori assumption that there is no phosphorus–carbon π-bonding in $P(C_6H_5)_5$, the occupation of π-orbitals calculated for PCl_5 is approximately $0.06\,\pi$ electron per orbital compared with the value of 0.21 for PF_5.

References

1. *Inorganic Syntheses*, vol. I, p. 99.
2. BROCKWAY, L. O., and BEACH, J. Y., *J. Amer. Chem. Soc.* **60**, 1836 (1938).
3. SIEBERT, H., *Zeit. anorg. und all. Chem.* **265**, 303 (1951).
4. MOUREU, H., MAGAT, M., and WETROFF, G., *Compt. Rend.* **205**, 276 (1937).
5. WILMSHURST, J. K., and BERNSTEIN, H. J., *J. Chem. Phys.* **27**, 661 (1957).
6. CARLSON, G. L., *Spectrochim. Acta* **19**, 1291 (1963).
7. TAYLOR, M. J., and WOODWARD, L. A., *J. Chem. Soc.* 4670 (1963).
8. CONDRATE, R. A., and NAKAMOTO, K., *Bull. Chem. Soc. Jap.* **39**, 1108 (1966).
9. GRIFFITHS, J. E., CARTER, R. P. Jr., and HOLMES, R. R., *J. Chem. Phys.* **41**, 863 (1964).
10. MOUREU, H., MAGAT, M., and WETROFF, G., *Compt. Rend.* **205**, 545 (1937).
11. CLARK, D., POWELL, H. M., and WELLS, A. F., *J. Chem. Soc.* 642 (1942).
12. PAYNE, D. S., *J. Chem. Soc.* 1052 (1953).
13. GERDING, H., and HOUTGRAAF, H., *Rec. Trav. Chim.* **74**, 5 (1955).
14. WIEKER, W., and GRIMMER, A-R., *Z. Naturforsch.* **21**, 1103 (1966).
15. ANDREW, E. R., BRADBURY, A., EADES, R. G., and JENKS, G. J., *Nature* **188**, 1096 (1960).
16. ANDREW, E. R., and WYNN, V. T., *Proc. Roy. Soc. A*, **291**, 257 (1966).
17. DiLORENZO, J. V., and SCHNEIDER, R. F., *Inorg. Chem.* **6**, 766 (1967).
18. CHIHARA, H., NAKAMURA, N., and SEKI, S., *Bull. Chem. Soc. Jap.* **40**, 50 (1967).
19. LINNETT, J. W., and MELLISH, C. E., *Trans. Faraday Soc.* **50**, 665 (1954).
20. GILLESPIE, R. J., *Can. J. Chem.* **38**, 818 (1960).
21. DUFFEY, G. H., *J. Chem. Phys.* **17**, 196 (1949).
22. VAN DER VOORN, P. C., PURCELL, K. F., and DRAGO, R. S., *J. Chem. Phys.* **43**, 3457 (1965).
23. COTTON, F., *J. Chem. Phys.* **35**, 228 (1961).
24. COULSON, C. A., *Nature* **221**, 1106 (1969).
25. CHANDLER, G. S., *J. Chem. Phys.* **47**, 1192 (1967).
26. CRAIG, D. P., and ZAULI, C., *J. Chem. Phys.* **37**, 601 (1962).
27. GILLESPIE, R. J., *J. Chem. Soc.* 4672 (1963).
28. VAN DER VOORN, P. C., and DRAGO, R. S., *J. Amer. Chem. Soc.* **88**, 3255 (1966).
29. LETCHER, J. H., and VAN WAZER, J. R., *J. Chem. Phys.* **45**, 2926 (1966).

CHAPTER 9

PHOSPHONITRILIC CHLORIDES

Introduction

The phosphazenes, also known as phosphonitriles or phosphonitrilic compounds, are inorganic polymers containing phosphorus and nitrogen atoms linked together in rings or chains. The cyclic phosphazenes, of general formula $(NPX_2)_n$, have been extensively investigated. They are remarkable for the formation of rings containing a large number of atoms: in the case of the fluorides, individual compounds have been characterized where n varies from 3 to 17. This behaviour is unusual, because very large rings are generally less stable than chain molecules. The delocalization of electrons around the ring accounts certainly in part for their stability.

A large number of inorganic compounds containing phosphazene heterocyclic units has been synthesized by exploiting the variations possible in n and the nature of the exocyclic groups, X. These compounds show certain resemblances to aromatic compounds and many of the synthetic and separational techniques of organic chemistry have been applied to their study.

The trimeric phosphonitrilic chloride, $(PNCl_2)_3$, (systematic name: 1,3,5,2,4,6-triaza-triphosphorine 2,2,4,4,6,6-hexachloride) has been extensively studied and there is an increasing amount of data being published about higher polymers such as $(PNCl_2)_4$. These compounds are the most widely investigated inorganic cyclic systems and the problems of their electronic structure have stimulated several theoretical accounts of the bonding.

Preparation

Phosphazenes were first obtained by Liebig[1] as the result of his attempt to prepare the amides of phosphonic acid by the reaction between PCl_5 and gaseous ammonia. The major product was phospham, (PN_2H), but Liebig also isolated a small quantity of a very stable substance containing nitrogen, phosphorus and chlorine which could be steam-distilled or boiled with acids or alkalis without appreciable decomposition. The empirical composition of this was established as $PNCl_2$[2,3] and vapour density measurements[4,5] showed the molecular formula was $(PNCl_2)_3$.

In order to prevent the formation of completely ammoniated products, Stokes[6] used ammonium chloride as the source of ammonia and studied its reaction with PCl_5 in a sealed tube between 150° and 200°:

$$3PCl_3 + 3NH_4Cl \rightarrow (PNCl_2)_3 + 12HCl$$

High pressures developed during the reaction due to the formation of HCl and the tubes had to be opened frequently to release the gas. The procedure was costly and dangerous because the tubes often exploded. The reaction product consisted of an amorphous mass which also contained crystalline material. $(PNCl_2)_3$ and $(PNCl_2)_4$ were extracted by petroleum ether and separated by fractional recrystallization from benzene. Higher polymers, identified as pentamer, hexamer and an oily substance of molecular weight corresponding to $(PNCl_2)_{10-11}$, were also obtained from the petroleum ether extract.

Stokes proposed the cyclic structure (I) for $(PNCl_2)_3$

and this has been fully substantiated by its chemical reactions and by structural investigations. $(PNCl_2)_3$ is the first member of a homologous series of heterocyclic compounds, of general formula $(PNCl_2)_n$ where $n = 3$ to 11. For example, $(PNCl_2)_4$ has the cyclic structure (II).

A further improvement in Stokes' preparative method was achieved by Schenck and Römer[7] who carried out the reaction in an autoclave. They confirmed an earlier observation that the optimum yield of phosphazenes is obtained by a reaction temperature of 120°. The most effective way of maintaining such a temperature is to use an inert solvent, such as *sym*-tetrachloroethane, as the medium for performing the reaction. This acts as a solvent for PCl_5 and allows a close control of the temperature of the reaction mixture. This procedure forms the basis of the preparative method generally used in the laboratory.[8]

The temperature of the reaction mixture is maintained between 128° and 143°C for 20 hr but continued heating is necessary to drive off dissolved hydrogen chloride. The reaction mixture is cooled, the solvent distilled off under reduced pressure and the residue fractionated *in vacuo* to separate the trimer and tetramer. At a pressure of $1 \cdot 33$ kNm^{-2}, the trimer distils between 120° and 128° and the tetramer between 180° and 187°C. The trimer is recrystallized from diethyl ether to give white crystals, m.p. $= 114$°C. The tetramer is recrystallized from benzene and the pure compound melts at $123 \cdot 5$°C.

The influence of varying experimental conditions on the products of reaction between PCl_5 and NH_4Cl in *sym*-tetrachloroethane has been studied thoroughly.[9] It has been established that, in addition to the cyclic polymers such as $(PNCl_2)_3$ and $(PNCl_2)_4$, a series of linear polymers is also obtained. The cyclic polymers are soluble in petroleum ether whereas the linear polymers are not. The individual cyclic polymers, from trimer to octamer inclusive, can be isolated by a combination of fractional extraction, distillation *in vacuo* and crystallization. The linear polymers remain as a brown, viscous oil of composition $(PNCl_2)_n \cdot PCl_5$ (where $n > 10$). The yield of $(PNCl_2)_3$ is greatly increased and the amount of linear polymers formed is reduced virtually to zero by a modification of the experimental procedure in which the PCl_5 is added slowly over a period of 6 to 8 hr to the refluxing suspension of NH_4Cl in *sym*-tetrachloroethane. Without this modification, about 27% of the product consists of linear polymers. With it, the products are: trimer, 63%; tetramer, 13%; higher cyclic polymers, 22%; and linear polymers, 2%. Each product is characterized by its infrared spectrum which serves as a means for its quantitative analysis.

The time of reaction between PCl_5 in solution and ammonium chloride is considerably lowered by the introduction of a small amount of the chloride of metals like Co, Mn, Zn, Cu, etc., which can form a coordination complex with ammonia. Thus only 2·5 to 3 hr is needed for the reaction to occur in *o*-dichlorobenzene in the presence of cobalt(II) chloride.

Phosphonitrilic chlorides have an aromatic-like odour. The vapour acts as an irritant to human tissues and causes inflammation of the eyes and lungs. The compounds must therefore be prepared and handled in an efficient fume cupboard.

Reactions of $(PNCl_2)_3$ and $(PNCl_2)_4$

All six chlorines in $(PNCl_2)_3$ can be progressively replaced by other groups. The replacement may be geminal (at a phosphorus bound to one chlorine) or non-geminal (at a phosphorus bound to two chlorines).

For example, geminal replacement occurs when phenyl groups are introduced into the phosphazene molecule by the Friedel–Crafts reaction. The reaction between $(PNCl_2)_3$ and benzene under reflux in the presence of aluminium chloride gives 2,2-diphenyl-4,4,6,6-tetrachlorocyclotriphosphazene(III). When the reflux is carried out for longer times, moderate yields of 2,2,4,4-tetraphenyl-6,6-dichlorocyclotriphosphazene(IV) are obtained. The geminal configuration of the phenyl groups in (III) and (IV) was established by hydrolysis to diphenylphosphinic acid, $(Ph_2)P{<}^{NH}_{OH}$. Hexaphenylcyclotriphosphazene(V) is obtained by reaction of $(PNCl_2)_3$ with benzene and aluminium chloride in an autoclave.

Mono-, tri- and pentaphenylphosphazenes have not been isolated and it appears that the substitution of one chlorine by a phenyl group facilitates the replacement of the second chlorine bound to the same phosphorus.

(V) has also been obtained by the reaction of phenyl magnesium bromide with $(PNCl_2)_3$. From $(PNCl_2)_4$, two tetraphenyl derivatives, two octaphenyl derivatives and two phenyl-substituted cleavage products are obtained. The tetraphenyl compounds are probably position isomers arising from the different possible arrangements resulting from the replacement of four out of eight chlorines. The octaphenyl derivatives are probably conformational isomers.

Non-geminal replacement occurs in reactions involving ammonia and some amines. In $(PNCl_2)_3$, the ring is approximately planar and, because of the relative disposition of the two chlorines bound to one phosphorus above and below the plane, non-geminal replacement can result in the formation of positional or geometrical isomers. The positional isomers of *bis*(amino)*tetrakis*(methylamino)cyclotriphosphazene (VIA and VIB) have been made by the reaction of $(PNCl_2)_3$ with ammonia and methylamine. Geometrical isomers of (VIA) can exist, depending on whether the NH_2 groups are on the same (*cis*) or opposite (*trans*) side of the planar ring.

(VIA)   (VIB)

In this compound, further isomerism can arise because of the different stereochemical arrangements possible for methylamino groups.

The reaction of $(PNCl_2)_3$ with ammonia produces the diamino derivative, $P_3N_3Cl_4(NH_2)_2$. Complete replacement of chlorine by amino groups occurs when $(PNCl_2)_3$ reacts with liquid ammonia over a period of 1–2 months. The analogous compound, octa-aminocyclotetraphosphazene has been made from $(PNCl_2)_4$.

$(PNCl_2)_3$ is comparatively stable towards hydrolysis. The reaction is slow enough to permit steam-distillation without appreciable decomposition. This contrasts sharply with the extremely rapid hydrolysis of P–Cl bonds in compounds like PCl_5 and $POCl_3$. An ether solution of $(PNCl_2)_3$ is hydrolysed by aqueous sodium acetate to the tetrahydrate of the trisodium salt of the trihydroxotrioxophosphazene (VII). The formation of three $P=O$ bonds removes the double bonding in the ring of $(PNCl_2)_3$.

When an aqueous solution of the sodium salt of (VII) is treated with a strong acid, breakdown of the ring structure occurs to form phosphoric acid and ammonia.

Hydrolysis of $(PNCl_2)_4$ occurs more readily than that of $(PNCl_2)_3$. A similar hydroxooxophosphazane, $[PNH(OOH)]_4$, is formed. This is a stable compound and quite resistent to further acid hydrolysis. It breaks down slowly into ammonia and phosphoric acid.

The reaction of alcohols with $(PNCl_2)_3$ in the presence of pyridine leads to hexa-alkyl esters of general formula (VIII). Again a similar fully-alkylated product has been made from $(PNCl_2)_4$.

The cyclic chlorophosphazenes when heated to 250–350°C polymerize to a high molecular-weight elastomer which, in its physical properties, resembles unvulcanized rubber.

It is stable at room temperature, but hardens and breaks up in damp air and swells in contact with benzene and other organic solvents. The molecular weight has been estimated to be within the range 37,000 to 173,000 and the elastomer is believed to consist of chains not cross-linked to one another. Depolymerization to a mixture of lower polymers begins at 350°C.

(VIII)

Many uses for chlorophosphazenes and their derivatives have been proposed including adhesives, additives in lubricants, plasticizers and in the flame-proofing of fabrics. In association with other materials, further uses have been described. For example, chlorophosphazenes and asbestos give semiconductor materials suitable for use as high-temperature gaskets, brake linings and electrical insulators.

Structure of $(PNCl_2)_3$

The electron diffraction study of the vapour[10] has confirmed Stokes' original proposal of a ring structure containing alternate phosphorus and nitrogen atoms, with two chlorine atoms bound to each phosphorus. The molecular paramaters are: P,N $= 0.165 \pm 0.004$ nm; P,Cl $= 0.197 \pm 0.003$ nm; Cl–P–Cl $= 107$–$110°$; and P–N–P $= 120° \pm 3°$. These values are based on a molecular model in which all the phosphorus and nitrogen atoms are assumed to be coplanar.

The structure of the solid has been determined by X-ray analysis.[11] The experimental data were subjected to a two-dimensional Fourier analysis and a three-dimensional least squares refinement. It was concluded from this work that the molecule consists of an almost flat ring of alternate phosphorus and nitrogen atoms, the bond lengths within the ring being between 0.157 and 0.161 nm and considered, to a first approximation, to be equal to one another. The bond angles within the ring have values between 118.5° and 121°. The P,Cl bond lengths are all identical (0.1975 nm), the planes containing each Cl–P–Cl group are perpendicular to the plane of the ring and the Cl–P–Cl bond angles are 101.77° and 102.05° (significantly less than the tetrahedral value).

The most significant features of this description are the shortness and equality of the P,N bonds, the approximate planarity of the ring and the equality of bond angles within it. The bond order is considerably greater than one because the bond length is very much shorter than the value, 0.178 nm, for a single bond between phosphorus and nitrogen (as found in the phosphoramidate ion, $NH_2PO_3^{2-}$). All these features suggest there must be extensive electron delocalization in the phosphazene ring.

A complete three-dimensional Fourier analysis[12] has more recently shown that the $(PNCl_2)_3$ ring is actually puckered and has a chair conformation.

Structure of $(PNCl_2)_4$

Two crystalline forms of the tetramer are known. Crystallization from solvents near room temperature gives the K-modification. At temperatures above 60°C, this is transformed irreversibly into the T-form. This second modification can also be obtained by crystallization from the melt.

Two X-ray diffraction analyses[13,14] have shown that K-$(PNCl_2)_4$ contains 'boat'-shaped molecules (Fig. 9.1). The unit cell is tetragonal and belongs to the space group P_{4_2}/n. The cell dimensions are: $a = b = 1·0844 \pm 0·0002$ nm; $c = 0·5961 \pm 0·0005$ nm. The P,Cl bond lengths have an average value of 0·1989 nm and are equal within experimental error. The P,N bond length reported in the more recent analysis[14] is 0·157 nm, compared with the value of 0·167 nm reported earlier.[13]. The bonds from the phosphorus atom point to the corners of a distorted tetrahedron. Bond angles are: P–N–P = 131·3° and N–P–N = 121·2°.

The T-form of $(PNCl_2)_4$ also crystallizes[15] in the tetragonal space group P_{4_2}/n and the unit cell dimensions are: $a = b = 1·5324$ nm; $c = 0·5988$ nm. The molecule is 'chair'-shaped (Fig. 9.1). The P,N bonds are of equal length (0·156 nm) and so are the P,Cl bonds (0·199 nm). Two different N–P–N angles are found, respectively 119·3° and 121·7°, and there are two distinct P–N–P angles as well, respectively 137·6° and 133·6°.

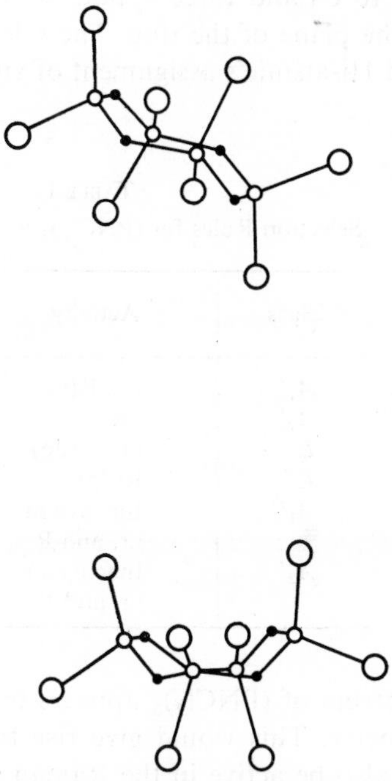

FIG. 9.1. The molecular shapes of T-$(PNCl_2)_4$ and K-$(PNCl_2)_4$.

The existence of two different forms of $(PNCl_2)_4$ indicates some degree of flexibility of the eight-membered ring. This is further shown by the vibrational spectrum of the tetramer in different physical states which has been interpreted in terms of molecules of various symmetries[16] (see below). The bond lengths within the tetramer are very close to those of the trimer and point to similar electron delocalization.

Vibrational Spectra of $(PNCl_2)_3$ and $(PNCl_2)_4$

The assignment of bands in the vibrational spectrum of $(PNCl_2)_3$ has been made on the basis of a planar molecule of D_{3h} symmetry.[17,18] D_{3h} symmetry elements (Fig. 9.2) are composed of a C_3 axis through the centre of the ring and perpendicular to it, three

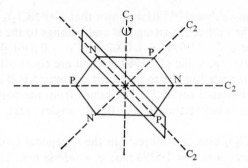

FIG. 9.2. Symmetry elements for $(PNCl_2)_3$ in the point group D_{3h}. Only one σ_v plane of symmetry is shown.

C_2 axes perpendicular to C_3 and three σ_v and one σ_h plane of symmetry, respectively perpendicular to and in the plane of the ring. The selection rules for this symmetry are summarized in Table I and Hisatsune's assignment of vibrational frequencies in the solid phase is given in Table II.

TABLE I

Selection Rules for $(PNCl_2)_3$ with D_{3h} Symmetry

Class	Activity	Number of vibrations
A_1'	i.r., R(p)	4
A_2''	i.r.	3
E'	i.r., R(dp)	6
E''	R(dp)	4
A_1''	Inactive in i.r. and R	1
A_2'	Inactive in i.r. and R	2

The vibrational spectrum of $(PNCl_2)_3$ appears to rule out a possible alternative 'chair' molecule of C_{3v} symmetry. This would give rise to seven polarized Raman lines and all infrared bands would also be active in the Raman spectrum. Only four polarized Raman lines are in fact observed so D_{3h} must be the correct symmetry.

The strong peak around 1210 cm^{-1} dominates the spectrum of $(PNCl_2)_3$. This is a degenerate stretching vibration of the whole ring. The frequency of this peak varies with ring size and it is located at 1315 cm^{-1} in $(PNCl_2)_4$. It is also changed when chlorine is replaced by a group or atom of different electronegativity, for example, in $(PNBr_2)_4$, it is found at 1175 cm^{-1}.

In the case of $(PNCl_2)_4$, several measurements of the infrared and Raman spectra have been carried out[19-23] and the data have been interpreted according to various molecular symmetries. The most recent treatment is by Hisatsune[16] and he correlates results on $(PNCl_2)_4$ in different phases with the symmetries D_{4h}, D_{2d}, S_4 and C_{2h} in the vapour, liquid and the two solid phases respectively. One major difficulty in the interpretation of the vibrational spectrum is that the frequencies observed may well depend on the sampling techniques because of the flexibility of the molecule and the ease with which it assumes

TABLE II

Assignment of Vibrational Frequencies for $(PNCl_2)_3$ in the Solid State[18]

Class	Approx. description of vibration	Number of vibration	Frequency (cm^{-1})
A_1'	Ring stretch	ν_1	783
	Ring deformation (in-plane)	ν_2	668
	PCl_2 sym. stretch	ν_3	365
	PCl_2 sym. bending	ν_4	304
A_2'	Ring stretch	ν_5	—
	PCl_2 wagging	ν_6	—
E'	Ring stretch	ν_7	1210
	Ring stretch	ν_8	1190
	Ring deformation (in-plane)	ν_9	524
	PCl_2 sym. stretch	ν_{10}	338
	PCl_2 wagging	ν_{11}	220
	PCl_2 bending	ν_{12}	140
A_1''	PCl_2 twisting	ν_{13}	—
A_2''	Ring deformation (out-of-plane)	ν_{14}	598
	PCl_2 asym. stretch	ν_{15}	540
	PCl_2 rocking	ν_{16}	189
E''	PCl_2 asym. stretch	ν_{17}	576
	Ring deformation (out-of-plane)	ν_{18}	204
	PCl_2 twisting	ν_{19}	177
	PCl_2 rocking	ν_{20}	162

various shapes. The discrepancies which exist between the sets of reported data could reflect real differences in molecular structures rather than experimental errors. For this reason, it is doubtful if a unique assignment is possible for complex molecules like the tetramer.

N.Q.R. Spectra

The trimer and tetramer have been examined in the solid state[24-28] by measuring the resonance originating from the quadrupolar nucleus ^{35}Cl, for which $I = 3/2$. The charge distribution in this nucleus is elliptical and there are $(2I + 1)$ different orientations possible relative to the electrical field of the orbital electrons. Each orientation is characterized by a nuclear magnetic quantum number, M, which has values, $I, (I - 1), \ldots, (1 - I), -I$. In the chloride ion, the electric field due to the orbital electrons is symmetric about the nucleus and all orientations are degenerate. When the chlorine atom forms a covalent bond, an asymmetric field is established. The asymmetry is confined to the bond axis and the field remains axially symmetric around this.

The energy, E_M, of a quadrupole state depends on Q, the quadrupole moment, and on q, the gradient of the electric field along the bond axis. For an axially symmetric field,

$$E_M = eQq \frac{[3M^2 - I(I + 1)]}{4 I(2I - 1)}$$

For ^{35}Cl, when $M = 3/2$ (and $-3/2$), $E_{3/2} = +eQq/4$: when $M = \frac{1}{2}$ (and $-\frac{1}{2}$), $E_{\frac{1}{2}} = -eQq/4$. Thus there are two doubly degenerate energy states and the transition between them should be observed as a single line in the N.Q.R. spectrum.

The spectrum of $(PNCl_2)_3$ actually shows four lines.[24,25] The frequencies reported by Negita and Satou[24] are at 27·630, 27·704, 27·834 and 27·903 MHz at 14°C and their observed intensities are respectively in the ratios 2:1:1:2: The electric field gradients for all chlorine nuclei in this molecule are evidently not exactly equivalent, and it is probable that the more intense lines are each due to two equivalent chlorines and the weaker lines each due to one chlorine in the molecule. The spectrum is therefore consistent with the puckered ring molecule in the 'chair' conformation shown by X-ray analysis. If the molecule were accurately planar, then all chlorines would be equivalent and only one N.Q.R. line would be observed.

The N.Q.R. spectrum of ^{35}Cl in $(PNCl_2)_4$ provided the first evidence of configurational isomerism in this molecule. The K-isomer shows[25,26] two resonance lines, located by Whitehead[26] at 28·0251 and 28·1565 MHz at 20°C and the T-isomer shows[24] four resonance lines at 27·265, 28·124, 28·182 and 28·616 MHz at 14°C. By observing the changeover from the N.Q.R. spectrum of one isomer to that of the other, it has been possible[28] to locate the transition temperature for $K \rightarrow T$ at 63 ± 1·5°C.

T-$(PNCl_2)_4$ evidently contains four kinds of non-equivalent chlorine atoms. The X-ray diffraction data do not indicate any differentiation between the chlorines on the basis of P,Cl bond lengths or Cl–P–Cl angles. It is possible, however, to distinguish between chlorine atoms on the basis of molecular symmetry. The 'chair'-shaped molecule has a two-fold axis which passes through two of the phosphorus atoms, P_2 and P_2', and the centre of symmetry. Figure 9.3 shows the molecule when viewed along this axis, with P_2 and P_2' and all four nitrogens, N_1, N_1', N_2 and N_2', co-planar. The four ^{35}Cl N.Q.R. frequencies correspond with atoms labelled Cl_1 (and Cl_1'), Cl_2 (and Cl_2'), Cl_3 and Cl_4.

N.M.R. Spectra

The only naturally occurring phosphorus isotope, ^{31}P, has a nuclear spin of $\frac{1}{2}$. In $(PNCl_2)_3$ and all higher cyclic polymers, only one peak is observed in the ^{31}P N.M.R. spectrum.[29] This is a clear demonstration that all the phosphorus atoms in a given polymer are equivalent. For $(PNCl_2)_3$ in benzene, the chemical shift (defined as

$$\frac{10^6 (H_s - H_r)}{H_r}$$

where H_s and H_r are the magnetic fields required for resonance in the sample an dreference compound respectively) relative to 85% orthophosphoric acid as reference compound, is $-19 ± 1$ ppm. For $(PNCl_2)_4$, the chemical shift is $+7 ± 1$ ppm and for $(PNCl_2)_5$, it is $+17 ± 1$ ppm. Beyond the pentamer, the shift appears to be independent of ring size.

Two resonances are observed in the polymers $(PNCl_2)_n \cdot PCl_5$: a large peak at $-20 ± 1$ ppm and a small one at $+7 ± 1$ ppm. These figures confirm the presence of phosphorus atoms in two different environments. These polymers are usually regarded as approximately linear in structure. Their properties are very different from the cyclic polymers, for example, their thermal stability is low, they readily dissociate in polar solvents to lose PCl_5 and they are very reactive towards water.

Bonding

The bonding in $(PNCl_2)_3$ and $(PNCl_2)_4$ can be described in relation to the observed stereochemistries of these molecules. The nitrogen and phosphorus atoms each have five valence electrons. The bond angles at nitrogen in the ring in $(PNCl_2)_3$ are due to sp^2 hybrid orbitals, two of which form σ-bonds with adjacent phosphorus atoms and the third of which contains a lone pair of electrons. The fifth electron is in a p_z-orbital, perpendicular to the plane of the ring. Phosphorus can be regarded as forming four σ-bonds from approximately tetrahedral sp^3 hybrids. Its fifth valence electron is in a d_π-orbital and overlap with the p_z nitrogen orbital leads to π-bonding. In $(PNCl_2)_4$, this description is an oversimplification because the bond angle at nitrogen in the ring shows considerable difference from the predicted value of 120°. In almost all cyclic phosphazenes studied, this angle is significantly greater than 120°.

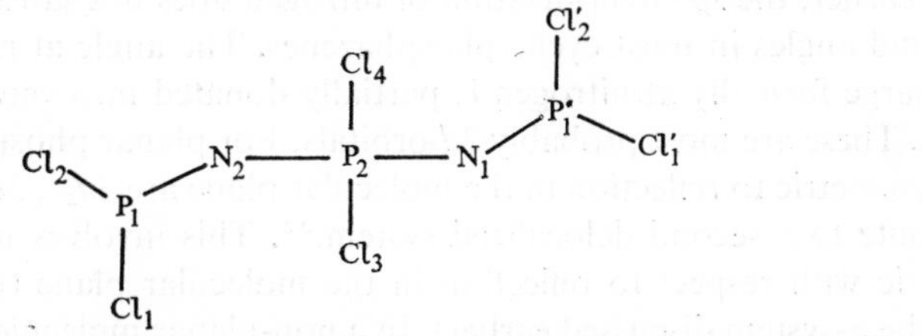

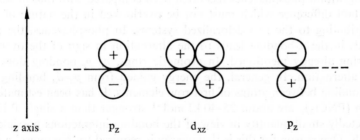

FIG. 9.3. The T-$(PNCl_2)_4$ molecule when viewed along the two-fold axis through P_2 and P_2' and the centre of symmetry.

FIG. 9.4. Interaction of d_{xz} and p_x orbitals in phosphazene rings.

The two d_π-orbitals which are involved in π-bonding are d_{xz} and d_{yz}. The interaction of d_{xz} with p_z-orbitals is illustrated in Fig. 9.3 and a π-orbital antisymmetric to reflection in the molecular plane is produced. The overlap can be extended to include the remaining atoms in the $(PNCl_2)_3$ ring with a resultant delocalization of the π-electrons. However, as shown in Fig. 9.4, matching-up the symmetry of the d_π-orbital necessitates an inversion in the sign of the p_z-orbitals on either side. In consequence, for a ring composed of three pairs of phosphorus and nitrogen atoms, it is impossible to match up completely the signs of the orbitals forming the π-bonds. In contrast, for a tetrameric phosphazene, the signs can be matched up throughout the ring and electron delocalization is complete.

These arguments have been used by Craig[30] to support the idea of aromatic-type binding in $(PNCl_2)_3$ and $(PNCl_2)_4$. He has proposed that p_z–d_{xz} interaction is more important than p_z–d_{yz}. The d_{yz}-orbital is destabilized by strong repulsions due to overlap with orbitals on the chlorine atoms and so it plays little part in the π-bonding. The d_{xz}-orbital is much less influenced by these atoms and it is more readily available for bonding with p_z-orbitals

4*

on nitrogen. A contrasting account of the π-bonding given by Dewar, Lucken and White-head[31] has suggested that it is not delocalized to the same extent as in benzene and can be represented in terms of three-centre P–N–P 'islands'. Further, they prefer equal contributions to the bonding from d_{xz}- and d_{yz}-orbitals and propose that linear combinations of these, designated $d\pi^a$ and $d\pi^b$, are involved.

$$d\pi^a = \frac{1}{\sqrt{2}}(d_{xz}+d_{yz}) \quad \text{and} \quad d\pi^b = \frac{1}{\sqrt{2}}(d_{xz}-d_{yz}).$$

Each of these combinations overlaps appreciably with the p_z-orbital of an adjacent nitrogen atom to form a three-centre molecular orbital which encloses two phosphorus atoms and one nitrogen. The two bonding electrons are effectively localized in this M.O. in a way which is formally analogous to the allyl cation.

As noted earlier, the sp^2 hybridization of nitrogen does not satisfactorily account for the observed bond angles in most cyclic phosphazenes. The angle at nitrogen may open if the lone-pair charge formally at nitrogen is partially donated into vacant acceptor orbitals on phosphorus. These are most probably $3d$-orbitals. For planar phosphazenes, the $3d$-orbitals which are symmetric to reflection in the molecular plane are $3d_{z^2}$, $3d_{x^2-y^2}$ and $3d_{xy}$ and these may contribute to a second delocalized system.[32] This involves molecular orbitals which are symmetric with respect to reflection in the molecular plane (termed π' to distinguish them from the π-system discussed earlier). In a non-planar molecule, both π- and π'-systems of bonding may still operate although further interactions between phosphorus and nitrogen orbitals must be taken into account.

The bonding within phosphazenes has often been compared with that in benzene. There is one important difference which must not be overlooked in the nature of the bonding orbitals contributing to the two delocalized systems. In phosphazenes, the participation of d_{xz} orbitals in the π-bonding leads to the alternation in sign of the overlap integrals around the ring already mentioned. In aromatic rings, p_π–p_π bonding does not involve this kind of alternation. In general, the energy gained from p_π–d_π bonding is less than from p_π–p_π bonding between atoms of first-row elements. It has been estimated[32] that the P,N bonds in $(PNCl_2)_3$ are about 25–40 kJ mol^{-1} stronger than a single P,N bond. This is an unexpectedly small quantity in view of the bonding interactions possible in addition to the σ-bonds. One reason for this is that energy is required to 'compress' a σ-bond from the length of a single bond to that appropriate for the bond with contributions from π-bonding. The neglect of this compression energy, which has been estimated to be as high as 125 kJ mol^{-1} in the case of benzene, could account for the apparently small gain in stability resulting from π-bonding in phosphazenes.

References

1. LIEBIG, J., *Ann.* **11**, 139 (1834).
2. GERHARDT, C., *Compt. Rend.* **22**, 858 (1846).
3. LAURENT, A., *Compt. Rend.* **31**, 349 (1850).
4. GLADSTONE, J. H., and HOLMES, J. D., *J. Chem. Soc.* **17**, 225 (1864).
5. GLADSTONE, J. H., and HOLMES, J. D., *Ann. Chim. Phys.* **3**, 465 (1864).
6. STOKES, H. N., *Amer. Chem. J.* **19**, 782 (1897).
7. SCHENCK, R., and RÖMER, G., *Ber.* **57B**, 1343 (1924).
8. *Inorganic Syntheses*, vol. VI, p. 94.
9. LUND, L. G., PADDOCK, N. L., PROCTOR, J. E., and SEARLE, H. T., *J. Chem. Soc.* 2542 (1960).

10. Brockway, L. O., and Bright, W. M., *J. Amer. Chem. Soc.* **65**, 1551 (1943).
11. Wilson, A., and Carroll, D. F., *J. Chem. Soc.* 2548 (1960).
12. Liquori, A. M., Pompa, F., and Ripamonti, A., *XVIIth International Congress of Pure and Applied Chemistry*, 1959, Paper A 1025.
13. Ketelaar, J. A. A., and de Vries, T. A., *Res. Trav. Chim.* **58**, 1081 (1939).
14. Hazekamp, R., Migchelson, T., and Vos, A., *Acta Cryst.* **15**, 539 (1962).
15. Wagner, A. J., and Vos, A., *Acta Cryst.* B **24**, 707 (1968).
16. Hisatsune, I., *Spectrochim. Acta* **25A**, 301 (1969).
17. Califano, S., *J. Inorg. Nucl. Chem.* **24**, 483 (1962).
18. Hisatsune, I., *Spectrochim. Acta* **21**, 1899 (1965).
19. Daasch, L. W., *J. Amer. Chem. Soc.* **76**, 3403 (1954).
20. Chapman, A. C., and Paddock, N. L., *J. Chem. Soc.* 635 (1962).
21. Stahlberg, U., and Steger, E., *Spectrochim. Acta* **23A**, 627 (1967).
22. Steger, E., and Stahlberg, U., *Zeit. anorg. und all. Chem.* **326**, 243 (1964).
23. Manley, T. R., and Williams, D. A., *Spectrochim. Acta* **23A**, 149 (1967).
24. Negita, H., and Satou, S., *Bull. Chem. Soc. Jap.* **29**, 426 (1956).
25. Torizuka, K., *J. Phys. Soc. Jap.* **11**, 84 (1956).
26. Whitehead, M. A., *Can. J. Chem.* **42**, 1212 (1964).
27. Kaplansky, M., and Whitehead, M. A., *Can. J. Chem.* **45**, 1669 (1967).
28. Dixon, M., Jenkins, H. D. B., Smith, J. A. S., and Tong, D. A., *Trans. Faraday Soc.* **63**, 2852 (1967).
29. Van Wazer, J. R., Callis, C. F., Shoolery, J. N., and Jones, R. C., *J. Amer. Chem. Soc.* **78**, 5715 (1956).
30. Craig, D. P., *J. Chem. Soc.* 997 (1959).
31. Dewar, M. J. S., Lucken, E. A. C., and Whitehead, M. A., *J. Chem. Soc.* 2423 (1960).
32. Craig, D. P., and Paddock, N. L., *J. Chem. Soc.* 4118 (1962).

CHAPTER 10

SULPHUR TETRAFLUORIDE

Introduction

Although the existence of SF_4 had been suspected for many years, it was not until 1950 that it was isolated and unequivocally identified for the first time. Since then, it has been widely studied. It has proved a valuable reagent for the preparation of new fluorocompounds and for the more convenient synthesis of many fluorocompounds already known.

The great chemical reactivity of SF_4 distinguishes it sharply from two other well-known fluorides of sulphur, SF_6 and S_2F_{10}. These are remarkably unreactive. Indeed, such is the inertness of SF_6 that it is used as an electrical insulator in high-voltage transformers and generators. In contrast, SF_4 is extremely reactive. For example, it is immediately hydrolysed to SOF_2 and HF on contact with water.

Halides of the general formula AX_4 are formed by several elements in Group VI. They are molecular compounds in which the symmetry is lower than tetrahedral. Their shapes have been related to a trigonal bipyramidal disposition of five electron pairs in the valence shell of A, four of the apical positions being occupied by bonding pairs to the halogens whilst the fifth is occupied by a lone pair of electrons. This arrangement requires the expansion of the valence shell of A to accommodate the extra electrons and presumably involves the use of one of its d-orbitals. Whether or not this description is valid can be discussed in the light of the experimental data now available on the exact shape and dimensions of the SF_4 molecule.

Preparation

The first reported attempt to make SF_4 was by the reaction between HF and S_4N_4 or SCl_4.[1] It appears to have been made in an impure state by the reaction of sulphur with CoF_3.[2] The compound was first isolated and some of its properties examined in 1950.[3] Carbon disulphide and fluorine react together to form $CF_3 \cdot SF_5$. When this is decomposed by the action of a high-voltage discharge, SF_4 is one of the products.

The most convenient laboratory-scale preparation is by the reaction of sulphur dichloride with excess sodium fluoride in acetonitrile solution.[4]

$$3SCl_2 + 4NaF \rightarrow SF_4 + S_2Cl_2 + 4NaCl$$

The reactants are heated under nitrogen at atmospheric pressure for 2 hr at 70–80°C. The principal impurity in the product is thionyl fluoride, SOF_2, formed by the adventitious hydrolysis of SF_4. SF_4 boils at $-38°$ and SOF_2 at $-43·7°C$ so their separation by distillation is difficult. Yields of up to 90% of SF_4, based on the SCl_2 taken, have been reported but the product, even after fractional distillation and collection of the fraction boiling between $-38°$ and $-37°C$, still contains around 5 mole % of SOF_2.

Better separations have been achieved by gas chromatography.[5] The most effective column packing is sodium fluoride coated with either tetramethylene sulphone or the dimethyl ether of tetraethylene glycol. A single passage through a preparative scale column of this material gives SF_4 of at least 98% purity.

Owing to the ease of hydrolysis of SF_4, the reactants must be completely dry and moisture must be rigorously excluded throughout the preparation. SF_4 is extremely toxic and this is an additional hazard in its preparation and use.

An ingenious chemical method for the purification of SF_4 exploits the formation of an adduct between SF_4 and a fluoride ion acceptor such as AsF_5. This compound has the ionic structure $SF_3^+AsF_6^-$. SOF_2 does not form such an adduct and can be pumped off as a gas at room temperature. Pure SF_4 is released from the adduct by treatment with SeF_4. This forms a stronger complex with AsF_5 and so displaces SF_4.

Reactions

The importance of SF_4 is due to its ability to act as a fluorinating agent. In general, it reacts with inorganic oxides or sulphides to replace oxygen or sulphur by fluorine. Its reaction with water occurs in two stages:

$$SF_4 + H_2O \rightarrow SOF_2 + 2HF$$
$$SOF_2 + H_2O \rightarrow SO_2 + 2HF$$

Iodine pentoxide, I_2O_5, reacts with SF_4 to give IF_5:

$$I_2O_5 + 5SF_4 \rightarrow 2IF_5 + 5SOF_2$$

Selenium tetrafluoride and the fluorides of heavy transition metals are most conveniently made by reaction of SF_4 with the appropriate oxides.[6] Complex fluorides have been synthesized by heating mixtures of alkali metal fluorides and transition metal oxides or sulphides with SF_4. Examples of such reactions are:

$$NaF + Nb_2O_5 \xrightarrow[350°C]{SF_4 at} NaNbF_6$$
$$NaF + WO_3 \xrightarrow[250°C]{SF_4 at} Na_2WF_8$$

SF_4 forms 1:1 adducts with other inorganic fluorides besides the compound with AsF_5 already mentioned. In principle, any fluoride that behaves as a Lewis acid can form an adduct, although the thermal stabilities of these compounds are very different. Of the adducts formed between SF_4 and SbF_5, IrF_5, BF_3, PF_5 and AsF_3, the most stable is $SbF_5 \cdot SF_4$ which melts without decomposition at 253°C and the least stable is $AsF_3 \cdot SF_4$ which shows appreciable dissociation above $-20°C$.

Strong inorganic oxidizing agents like CrO_3 and CeO_2 convert SF_4 to sulphur oxotetrafluoride, SOF_4.

Sulphur tetrafluoride is an important selective fluorinating agent towards organic compounds.[7] It reacts with the carbonyl group in aldehydes and ketones to give organodifluorides containing CF_2 groupings.

$$>C=O + SF_4 \rightarrow >C{<}^F_F + SOF_2.$$

In its reaction with a carboxylic acid, the carbonyl group is converted to a trifluoromethyl group. This provides a general method for the synthesis of compounds containing $-CF_3$ groups. For example, with benzoic acid:

$$C_6H_5COOH + SF_4 \rightarrow C_6H_5COF + HF + SOF_2$$
$$C_6H_5COF \quad + SF_4 \rightarrow C_6H_5CF_3 \quad + SOF_2$$

Olefinic and acetylenic bonds are not affected by SF_4 and so it is a selective fluorinating agent for unsaturated carboxylic acids:

$$CH_2=CHCOOH \xleftarrow{\ SF_4\ } CH_2=CHCF_3$$

The thiocarbonyl group is fluorinated by SF_4 in a similar manner. Free sulphur is a product of this reaction.

$$2>C=S + SF_4 \rightarrow 2>CF_2 + 3S$$

Vibrational Spectrum

There are several shapes possible for a molecule of formula AX_4. They include (1) tetrahedral (T_d symmetry), (2) square planar (D_{4h}), (3) trigonal pyramidal (C_{3v}), derived from a trigonal bipyramid with one of the polar positions unoccupied, and (4) distorted tetrahedral (C_{2v}), derived from a trigonal bipyramid with one of the equatorial positions unoccupied (Fig. 10.1). For each of these models, the nine vibrations in the molecule may be classified according to the various species in the particular symmetry group. The classes to which these vibrations belong and their Raman and infrared activities are summarized

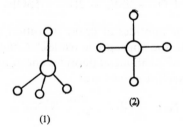

(1)   (2)

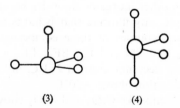

(3)   (4)

FIG. 10.1. Possible shapes for a AX_4 molecule.

in Table I. Provided the fundamental vibrations can be assigned without ambiguity, a definite conclusion on the shape of the SF_4 molecule should be possible from the analysis of its vibrational spectrum because the number of Raman-active and the number of infrared-active vibrations are different for each point group.

The infrared spectrum of SF_4 gas is complex because of the occurrence of many combination and overtone bands as well as the fundamentals. Nevertheless, Dodd, Woodward and Roberts[8] were able to assign at least five bands to fundamental vibrations. These were located at 532 (ν_7), 557 (ν_3), 728 (ν_8), 867 (ν_6) and 889 cm^{-1} (ν_1). Assignments were largely based on the intensities of these bands and their contribution to overtone and combination bands. This eliminated the possibility that SF_4 could be either square planar or tetrahedral and left either C_{2v} or C_{3v} as the correct symmetry. The Raman spectrum, measured on liquid SF_4 at $-60°C$ by the same authors, and the infrared spectrum of the gas were most satisfactorily interpreted on the basis of C_{2v} molecular symmetry. The remaining four fundamentals were assigned to bands at 235 (ν_4), 645 (ν_5), 715 (ν_2) and 463 cm^{-1} (ν_9). Some of these were rather speculative. Thus ν_4 and ν_5 were not directly observed but were deduced from combination bands. Their work remains an important contribution in establishing for the first time the shape of the SF_4 molecule and their conclusion about its symmetry has been fully justified in subsequent investigations of SF_4 by other techniques.

More recently another set of vibrational assignments has been published[9] (Table II). The infrared spectrum of SF_4 is shown in Fig. 10.2. The assignments for ν_4, ν_9 and ν_3 are

TABLE I

The Number of Fundamental Vibrations and Their Classes and Activities for Various Possible Structures of SF_4

Point group	No. of fundamental vibrations	No. of Raman-active vibrations	No. of infrared vibrations	No. of polarized Raman lines
T_d	4	$4(A_1, E, 2T_2)$	$2(2T_2)$	$1(A_1)$
D_{4h}	7	$3(A_{1g}, B_{1g}, B_{2g})$	$3(A_{2u}, E_u)$	1
C_{3v}	6	$6(3A_1, 3E)$	$6(3A_1, 3E)$	$3(3A_1)$
C_{2v}	9	$9(4A_1, A_2, 2B_1, 2B_2)$	$8(4A_1, 2B_1, 2B_2)$	$4(4A_1)$

TABLE II

Vibrational Assignments[9] for SF_4 (C_{2v} symmetry)*

Class	Vibration	Frequency (cm^{-1})	Approximate description
A_1	ν_1	891·5	Sym. S–F'' stretch
	ν_2	558·4	Sym. S–F' stretch
	ν_3	353	
	ν_4	171	
A_2	ν_5	645†	
B_1	ν_6	867	Antisym. S–F' stretch
	ν_7	532	
B_2	ν_8	728	Antisym. S–F'' stretch
	ν_9	226	

* The convention is adopted that B_1 vibrations are symmetric with respect to the xy plane. A_1, B_1 and B_2 vibrations involve dipole moment changes parallel to the z, x and y axes respectively.

† Assignment taken from ref. 8.

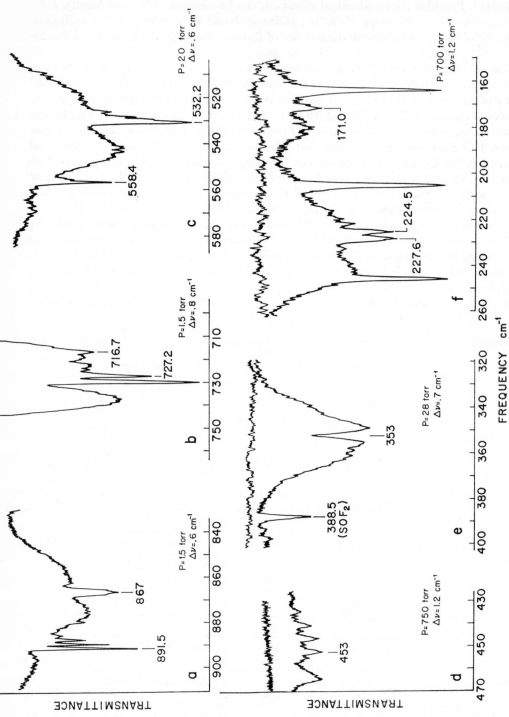

FIG. 10.2. The infrared spectrum of SF_4 in the gas phase* (reproduced with permission from Levin, I. W., and Berney, C. V., *J. Chem. Phys.* **44**, 2557 (1966)).

* P is the gas pressure in torr units (1 torr is equivalent to 133·322 Nm⁻²)

different and the band at 558·4 cm⁻¹ is identified with ν_2 instead of ν_3. In other respects, the assignments agree with the original ones. The bands at 463 and 715 cm⁻¹ assigned by Dodd and co-workers as fundamentals are considered to be overtones of the fundamentals at 226 and 353 cm⁻¹ respectively. Two impurities are present in the sample whose spectrum is shown in Fig. 10.2. They are SOF_2 (identified by a band at 388·5 cm⁻¹) and HF (three rotational lines at 163·9, 204·5 and 245·0 cm⁻¹).

The assignments in Table II have been substantiated by a normal coordinate analysis which shows general agreement between observed and calculated mean vibrational amplitudes. The calculated values of the force constants for $S-F_{eq.}$ and $S-F_{ax.}$ bonds are respectively 480 and 350 Nm⁻¹. This clearly shows that there are two sets of fluorine atoms in SF_4 which are differentiated by the strength of their bonds to sulphur.

Structure

The structure and dipole moment have been determined from the rotational spectrum of SF_4 measured in the microwave region.[10] The molecular parameters are given in Table III. The shape is illustrated in Fig. 10.3. Two of the S,F bonds, of length 0·1646 ± 0·0003 nm, are nearly collinear. These are the axial bonds. The other two, of length 0·1545 ± 0·0003 nm, lie in a plane at right angles to the plane containing the axial bonds. These are the equatorial bonds.

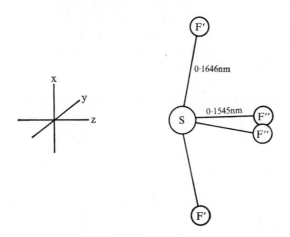

Fig. 10.3. The structure of SF_4 as determined by microwave spectroscopy.

The dipole moment has been calculated as 0·632 ± 0·003 D from observations of the splitting of lines in the rotational spectrum under the influence of an applied electric field (Stark effect).

There is very close agreement between the bond lengths and angles found spectroscopically and those calculated from electron diffraction work[11] (see Table III). The only difference of any significance is the slight increase in values of the bond angles determined by electron diffraction. Both investigations have fully confirmed the structural model of C_{2v} symmetry.

4a*

TABLE III

Molecular Parameters of Sulphur Tetrafluoride

	Bond lengths (nm)		Bond angles (deg.)	
	S, F′	S, F″	F′SF′	F″SF″
Microwave spectroscopy	0·1646	0·1545	173°4′	101°33′
Electron diffraction	0·1643	0·1542	176°48′	103°48′

N.M.R. Spectra

Fluorine compounds are well suited to study by N.M.R. spectroscopy because the ^{19}F nucleus has a high resonance sensitivity and the chemical shifts between flourines in non-equivalent environments are in general quite large.

At −98°C and at 30 MHz, the spectrum of SF_4 shows two triplets which are of equal intensity and widely separated from each other[12] (Fig. 10.4). An intensity ratio of 1:2:1 is observed for the lines comprising each triplet and there is equal hyperfine splitting in each triplet ($J = 78$ Hz). SOF_2 impurity appears as a single sharp peak. These features establish the presence of two different environments of the fluorine atoms, each containing the same number of nuclei. The symmetries D_{2d}, D_{2h}, C_{4v}, D_{4h} and T_d are thereby excluded because each would give a single resonance line only.

The triplet fine structure arises from spin–spin coupling between the two equivalent sets of fluorine atoms. The number of nuclear spin states of equivalent sets of nuclei is $2I + 1$, where $I =$ total spin of the set. In SF_4, the triplet fine structure shows $I = 1$. Since the nuclear spin for each fluorine is $\frac{1}{2}$, each equivalent set contains two fluorine atoms. The only possible model to fit these data is that of C_{2v} symmetry (4).

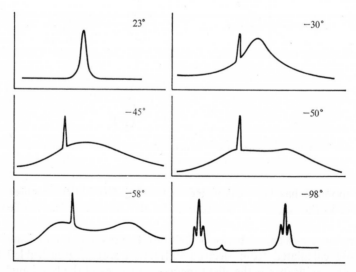

FIG. 10.4. The dependence with temperature of the ^{19}F nuclear magnetic resonance spectrum of SF_4 (reproduced with permission from Muetterties, E., and Phillips, W. D., *J. Amer. Chem. Soc.* **81**, 1084 (1959)).

The N.M.R. spectrum of SF_4 shows marked changes with temperature. When the temperature of the sample is raised from $-98°C$ the peak widths increase until they collapse into a single broad resonance at $-47°C$. The width of this decreases with further increase in temperature (Fig. 10.4). This behaviour is typical of fluorides of low symmetry where the exchange of atoms between the two non-equivalent fluorine environments is catalysed by fluoride impurity.

The order and mechanism of the exchange in SF_4 have not been precisely determined although kinetic measurements have shown it is of second or higher order. Fluorine exchange is also found in SbF_5 and it has been established that liquid SbF_5 is associated at least to dimers and probably to cyclic or linear polymers through fluorine bridges. This bridging could be the mechanism for fluorine exchange in SbF_5 and similarly in SF_4. Muetterties and Phillips[12] have suggested that a possible dimer of SF_4 which could act as an intermediate in fluorine exchange is:

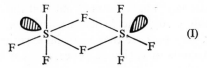

(I)

Under higher resolution, the ^{19}F spectrum shows further splitting. At 56.4 MHz and a field of 1.4 T., second-order splitting into doublets of all the peaks of both triplets is observed.[13]

Molecular Association of Sulphur Tetrafluoride

To seek evidence for the existence of a dimer, the infrared spectrum of SF_4 at very low temperatures has been examined by Redington and Berney.[14] The samples were in the form of a deposit of SF_4 trapped in an argon matrix at $\sim -269°C$ on a caesium iodide window and the spectrum was measured between 1400 and 400 cm^{-1}. Deposits with an Ar:SF_4 ratio of 400 were found to have spectra consisting of sharp peaks, sometimes split into multiplets, at frequencies somewhat lower than those of the corresponding gas phase bands. The sample was allowed to warm up to $-242°C$, cooled back to $-269°C$ and the spectrum measured again. The original peaks were found to have lower intensities and some new peaks had appeared. Repeated thermal cycling showed that these peaks grew at the expense of the original ones. For example, in the initial spectrum of the sample, a strong band at 707 cm^{-1} was observed due to the fundamental, ν_8. On thermal cycling, the intensity of this decreased while the weak absorption around 692 cm^{-1} in the initial spectrum increased markedly in intensity, a new band appeared at 669 cm^{-1} and there was a general increase in the extent of absorption below 700 cm^{-1}. These changes were believed to correspond with the formation of distinct associated species when the SF_4 molecules diffused through the matrix as its temperature was raised. Redington and Berney suggested a different dimeric species to account for their results. In this, two fluorine atoms of one SF_4 molecule are associated with the lone pair of electrons on the sulphur atom of a second SF_4 molecule:

(II)

Muetterties and Phillips[15] have criticized some of these conclusions and this proposed model of the dimer. The observed perturbation of the infrared spectra need not necessarily be due to bonding interactions: it could be induced by the matrix. They also point to the improbability of an associated species in which two fluorines from one SF_4 molecule act as bridging atoms to the second sulphur and yet a non-bonding pair of electrons is interposed between the two molecules. On theoretical grounds, one would expect bonding interactions to be a minimum and non-bonding repulsions to be a maximum in such a species.

The essential difference between the alternative models is that fluorine exchange in (I) would be an intermolecular process whereas in (II) it would be an intramolecular process. Muetterties and Phillips have commented on the absence of splitting due to ^{33}S–^{19}F coupling as evidence in favour of (I). ^{33}S possesses a nuclear spin of 3/2 and so a quartet of lines should result from those molecules which contain ^{33}S if there is no intermolecular fluorine exchange. Redington and Berney[16] have pointed out that the abundance of ^{33}S in naturally-occurring sulphur is only 0·76% and hence the detection of such splitting would be very difficult. Accordingly, they propose that a crucial experiment to enable a decision to be made between the two kinds of exchange process would be the measurement of the N.M.R. spectrum of SF_4 enriched with ^{33}S.

Bonding

The bonding in SF_4 has, like that in PCl_5, been described in terms of $sp^3d_{z^2}$ hybrids according to the Pauling–Slater theory of directed valence. The molecular shape is derived from a trigonal bipyramid in which one of the equatorially directed hybrid orbitals is occupied by a lone pair of electrons. Qualitative arguments based on the spatial arrangement of four bonding pairs and one lone pair of electrons so that repulsion between them is a minimum (Sidgwick–Powell principle) also lead to the same molecular shape.

On the assumption that d-orbitals participate in the bonding, it is possible to account in a simple way for the reactivity of SF_4 which contrasts so strongly with the chemical inertness of SF_6. In SF_4, where the configuration of the sulphur atom is either s^2p^3d or sp^4d, the promotion of one electron to give sp^3d^2 makes it possible for the sulphur to form new bonds with an attacking group. In the case of SF_6, no further promotion of valence-shell electrons is possible and attack at the sulphur atom to form an extra bond cannot occur. An additional factor responsible for the great differences between the two compounds is the complete shielding of the sulphur by fluorines in SF_6 which hinders the access of an attacking group whereas in SF_4, such access is easy due to the irregular molecular shape.

To attempt to account for the observed difference in the lengths of the equatorial and axial bonds, it has been proposed that two different sets of hybrid orbitals are involved in bonding. The two shorter equatorial bonds are formed from orbitals on sulphur which are intermediate between sp^2 hybrids and pure p-orbitals and the longer axial bonds from pd hybrid orbitals. This could account for the equatorial bond angle being between 90° and 120°.

The extent to which 3d-orbitals of the central atom are involved in compounds of the typical elements of the later groups in the Periodic Table has been widely discussed.[17,18] One view is that these orbitals are of only marginal significance in the bonding within molecules like PF_5, SF_4 and ClF_3. Indeed, it is possible that there is no involvement at all of the d-orbitals. Rundle[19] has described PF_5 in terms of conventional two-centre two-electron bonds (using the $3s$- and two of the $3p$-orbitals of phosphorus) in the equatorial plane with the two axial fluorines bound by a three-centre four-electron bond which involves only a p-orbital from each fluorine and the remaining p-orbital of phosphorus. This kind of description is applicable to SF_4, one of the equatorial positions being occupied by the lone pair. In both molecules one would expect the axial to be less strongly bound than the equatorial fluorines and hence the length of the axial bonds to be greater than that of the equatorial bond.

The three-centre four-electron bond has been a useful concept in describing the bonding in noble-gas compounds and there is no reason why such a bond should not be found much more generally in covalent molecules. According to Rundle, it provides a means whereby the coordination number of an atom like P, S or Cl can be increased without expansion of its valence shell.

This description is probably a more accurate representation of the bonding within SF_4 than the valence-bond model using sp^3d hybrids. However, a complete M.O. treatment must include a recognition of other contributions to the bonding, for example, from s- and d_{z^2}-orbitals in σ-bonding of the axial fluorines and from other d-orbitals of sulphur in π-bonding.

References

1. RUFF, O., and THIEL, C., *Chem. Ber.* **38**, 549 (1905).
2. FISCHER, J., and JAENCKNER, W., *Angew. Chem.* **42**, 810 (1929).
3. SILVEY, G. A., and CADY, G. H., *J. Amer. Chem. Soc.* **72**, 3624 (1950).
4. TULLOCH, C. W., FAWCETT, F. S., SMITH, W. C., and COFFMAN, D. D., *J. Amer. Chem. Soc.* **82**, 539 (1960).
5. ROBSON, J. W., and ASKEW, W. B., *J. Chromatog.* **7**, 409 (1962).
6. KEMMITT, R. D. W., and SHARP, D. W. A., *J. Chem. Soc.* 2496 (1961).
7. HASEK, W. R., SMITH, W. C., and ENGELHARDT, V. A., *J. Amer. Chem. Soc.* **82**, 543 (1960).
8. DODD, R. E., WOODWARD, L. A., and ROBERTS, H. L., *Trans. Faraday Soc.* **52**, 1052 (1956).
9. LEVIN, I. W., and BERNEY, C. V., *J. Chem. Phys.* **44**, 2557 (1966).
10. TOLLES, W. M., and GWINN, W. D., *J. Chem. Phys.* **36**, 1119 (1962).
11. KIMURA, K., and BAUER, S. H., *J. Chem. Phys.* **39**, 3172 (1963).
12. MUETTERTIES, E., and PHILLIPS, W. D., *J. Amer. Chem. Soc.* **81**, 1084 (1959).
13. BACON, J., GILLESPIE, R. J., and QUAIL, J. W., *Can. J. Chem.* **41**, 1016 (1963).
14. REDINGTON, R. L., and BERNEY, C. V., *J. Chem. Phys.* **43**, 2020 (1965).
15. MUETTERTIES, E., and PHILLIPS, W. D., *J. Chem. Phys.* **46**, 2861 (1967).
16. REDINGTON, R. L., and BERNEY, C. V., *J. Chem. Phys.* **46**, 2862 (1967).
17. BROWN, R. D., and PEEL, J. B., *Aust. J. Chem.* **21**, 2617 (1968).
18. MITCHELL, K. A. R., *Chem. Rev.* **69**, 157 (1969).
19. RUNDLE, R. E., *J. Amer. Chem. Soc.* **85**, 112 (1963).

CHAPTER 11

SULPHUR TRIOXIDE

Introduction

Sulphur trioxide is produced as an intermediate in the manufacture of sulphuric acid by the contact process at the rate of millions of tons annually. Although almost all of it is converted immediately into sulphuric acid or oleum there is an increasing demand for sulphur trioxide itself, particularly as a reagent for the sulphonation of organic chemicals, for example, in the production of synthetic detergents. In view of its industrial importance and widespread use, it is surprising that, even today, some of the physico-chemical properties of sulphur trioxide are still not completely understood.

Much of the complexity of the chemistry of sulphur trioxide is associated with the ease with which it polymerizes. In the solid state, at least four different forms have been described: α (high-melting and asbestos-like), β (low-melting and asbestos-like), γ (ice-like) and a gelatinous form. Liquid sulphur trioxide is stable when anhydrous but polymerizes in the presence of traces of water. The interrelation between the different forms of the solid has been established although detailed structural analyses have been carried out so far on β- and γ-SO_3 only.

Preparation

Sulphur trioxide was first recognized as the white sublimate formed in the thermal decomposition of metal sulphates such as those of copper and iron. It is also obtained on heating fuming sulphuric acid and is formed in small amounts during the combustion of sulphur in air.

The most important method of preparation is undoubtedly by the reaction of sulphur dioxide with oxygen.

$$2SO_2 + O_2 \rightleftharpoons 2SO_3$$

This reaction proceeds rapidly under the influence of catalysts such as platinum, platinized asbestos, ferric oxide or vanadium pentoxide. It is the basis of the contact process for the manufacture of sulphuric acid. Heat is evolved in the oxidation of sulphur dioxide and the yield of trioxide decreases the higher the temperature of the reaction mixture. Acid of a specific concentration can be made by dissolving sulphur trioxide in the appropriate quantity of water. In practice, it is found that the gas is absorbed with great difficulty by water or dilute acid and it is usual to absorb it in concentrated sulphuric acid (97–98% by weight).

Reactions

In many of its reactions with inorganic compounds, sulphur trioxide acts as a strong oxidizing agent, being itself reduced to sulphur dioxide. It converts many metal compounds into sulphates and serves as a reactant for the preparation of various oxo-sulphur compounds.

Alkali metal chlorides absorb gaseous sulphur trioxide with the formation of compounds of general formula $M(SO_3)_nCl$. Absorption by KCl at $0°C$ has been used to separate sulphur trioxide vapour from other components of gas mixtures. On heating $M(SO_3)_nCl$, some of the sulphur trioxide is evolved, then chlorine and sulphur dioxide and the alkali metal sulphate remains behind. The absorption of sulphur trioxide by CaF_2 gives $Ca(S_2O_6F)_2$. When this is heated, the compound $S_2O_5F_2$ is released.

Chlorosulphonic acid is formed by reaction with HCl whereas HBr and HI are oxidized respectively to bromine and iodine. Reaction with CCl_4 proceeds thus:

$$2SO_3 + CCl_4 \rightarrow COCl_2 + S_2O_5Cl_2$$

Metal sulphides are oxidized to the corresponding sulphate. With boron trichloride, the reaction is:

$$2BCl_3 + 4SO_3 \rightarrow 3SO_2Cl_2 + B_2O_3 \cdot SO_3$$

Polysulphuric acids such as $H_2S_2O_7$ and $H_2S_3O_{10}$ are produced by reaction between sulphur trioxide and H_2SO_4. Evidence for their formation has been drawn from Raman[1] and infrared[2] spectroscopic studies on solutions of SO_3 in H_2SO_4. There is some doubt over the formation of significant amounts of any higher polymer such as $H_2S_4O_{10}$.

Sulphur trioxide is highly reactive towards many organic compounds. Three types of reaction occur:[3] (i) sulphonation (formation of a C–S bond), e.g. reaction with benzene which produces mainly benzene sulphonic acid, $C_6H_5SO_3H$, and some diphenyl sulphone, $C_6H_5SO_2C_6H_5$; (ii) sulphation (formation of an O–S bond), e.g. reaction with methanol to form methyl hydrogen sulphate, CH_3OSO_3H; (iii) sulphamation (formation of a N–S bond), e.g. the formation of $(C_2H_5)_2NSO_3H$ from diethylamine.

Many organic substances form complexes with SO_3. Those with pyridine, dioxane and trimethylamine are probably the best known and are widely used in synthetic work to moderate the high reactivity of SO_3. Sulphur trioxide is a Lewis acid and the compounds it complexes with are Lewis bases (electron donors). Thus $(CH_3)_3N$ reacts in the vapour phase or in solvents like SO_2 or $CHCl_3$ with SO_3 to give a 1:1 complex, $O_3S \leftarrow N(CH_3)_3$. This is a stable solid, m.p. $239°C$ (with decomposition). It is soluble to some extent in water and its aqueous solutions undergo hydrolysis only slowly. These can therefore be conveniently used for the sulphation or sulphamation of many organic compounds.

Similar 1:1 adducts between pyridine or dioxane and SO_3 are known. The complex with dioxane has been recommended as a better sulphonating agent for polystyrene than either sulphuric or chlorosulphonic acid. The pyridine and dioxane complexes are, in general, more reactive than the trimethylamine complex because of the greater basic strength and consequent stronger complexing ability of trimethylamine compared with either of the other two molecules.

Structure

The tantalizing problems concerning the structure of the various solid forms of sulphur trioxide have still not been completely solved.

In 1924 and 1926,[4,5] Smits and Schoenmaker described their investigations on three modifications, identified by the suffixes α, β and γ. The ice-like form, now referred to as γ-SO_3 but called α-SO_3 by Smits and Schoenmaker, is made by crystallization of the thoroughly-dried liquid. It melts sharply at 16·8°C. α-SO_3 is a form resembling asbestos in appearance and can be made by repeated distillations of anhydrous γ-SO_3, the vapour being condensed each time at liquid nitrogen temperatures. α-SO_3 can be melted under pressure at temperatures greater than 80°C. It is characterized by very low vapour pressure at room temperature and Smits and Schoenmaker found, on partial distillation above room temperature, that the vapour pressure of the solid remaining became even lower. They concluded this behaviour was that of mixed crystals containing at least two components differing widely in vapour pressure. β-SO_3 is another asbestos-like form which is normally produced when SO_3 is stored in the solid state or at temperatures not far above 16·8°C. It melts slowly between 30° and 40 °C provided polymerization is not far advanced. Small amounts of water appear to be essential for the formation of this modification. Smits and Schoenmaker[6] attempted to differentiate between the three forms by X-ray powder photography but this proved impossible because all gave the same pattern. This suggests that, under their experimental conditions, exposure to X-rays caused the metastable forms to change to the stable one.

After this discouraging result, interest in the structure of solid SO_3 waned for several years until proposed structures for γ- and β-SO_3 were published[7] based on their Raman spectra. γ-SO_3 was believed to consist largely of cyclic S_3O_9 molecules and β-SO_3 of long chains of alternate sulphur and oxygen atoms.

The crystal structure of γ-SO_3 was determined in 1941 by X-ray diffraction.[8] Single crystals were obtained by inducing crystallization in the supercooled liquid. γ-SO_3 is ortho-rhombic and the unit cell, containing 12 SO_3 units, has the dimensions: $a = 1·23$ nm, $b = 1·07$ nm, $c = 0·53$ nm. The structure previously proposed was substantiated. The crystal has a molecular lattice built up of S_3O_9 molecules. These are puckered rings of C_{3v} symmetry, each being composed of three SO_4 tetrahedra, two oxygens from each tetrahedron being shared with the other sulphurs to form a six-membered ring. The length of the S,O bond within the ring was 0·16 nm and that of the other S,O bonds was 0·14 nm.

The molecular dimensions and the valence angles have been more precisely determined recently. Figure 11.1 shows the shape of the S_3O_9 ring as calculated[9] from the experimental data of Pascard and Pascard-Billy.[10] The S,O bond lengths within the ring are 0·1626 ±0·0007 nm and the other bond lengths outside the ring are: axial, 0·1371 ± 0·0013 nm; equatorial, 0·143 ± 0·0013 nm. The lengths of S,O bonds in many compounds have been measured and their values range from 0·166 nm in SF_5OOSF_5 through 0·149 nm in SO_4^{2-} and 0·1405 nm in SO_2F_2 to the short axial bond length in S_3O_9. Cruickshank[11] has described the shortening of bonds in SO_4^{2-}, as compared with a theoretical single bond length of 0·169 nm, in terms of a contribution to the bonding from π-bonds. In addition to the four tetrahedrally-disposed σ-bonds, π-bonds result from overlap of two p-orbitals on oxygen and the $3d_{x^2-y^2}$ and $3d_{z^2}$ orbitals on sulphur. From the viewpoint of the valence-bond theory, each S,O bond in SO_4^{2-} is regarded as having a π-bond order of $\frac{1}{2}$. On this basis, the bonds outside the S_3O_9 ring must have considerable double bond character and those in the ring only a relatively small amount. The difference in length between bonds

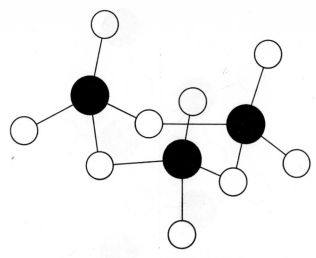

FIG. 11.1. The shape of the S_3O_9 molecule.

inside and outside the ring is to be expected in view of the greater extent of π-bonding when an oxygen atom is linked to one rather than two sulphur atoms. The difference of 0·006 nm between the axial and equatorial bonds is not and it is difficult to explain why the axial S,O bond, believed to be the shortest on record between these atoms, should be significantly shorter than the equatorial S,O bond. The axial oxygens of neighbouring tetrahedra are 0·297 nm apart compared with a S,S separation of 0·282 nm. This means the axial bonds are not exactly parallel but diverge slightly.

The crystal structure of β-SO_3 has been determined.[12] Single crystals of this modification are difficult to obtain and the X-ray crystallographic analysis was actually performed on a crystal taken from a sealed bulb containing sulphur trioxide, in which good crystals of β-SO_3 had developed over a period of some 25 years. β-SO_3 is composed of chains parallel to the needle axis. These are formed by SO_4 tetrahedra linked so that a spiral chain of –O–S–O–S–O–S–O– atoms results. Each chain has a two-fold screw symmetry axis. The structure is illustrated in Fig. 11.2 and interatomic distances (nm) are: S...S, 0·285; S,O_1, 0·163; S,O_1, 0·159; S,O_2, 0·141; and S,O_3, 0·141. In the crystal the oxygen atoms form a slightly distorted cubic close-packed structure.

There is a formal structural analogy between γ-SO_3 and silicates of the benitoite type and between β-SO_3 and the pyroxenes. However, rings and chains of sulphur trioxide have configurations appreciably different from those of silicates. Thus, in both β- and γ-SO_3, the valence angle at the oxygen atom connecting two tetrahedra is significantly smaller than in the corresponding silicates.

Further facts about the nature of α- and β-SO_3 have now been established. α-SO_3, made by the condensation of SO_3 vapour on a surface cooled below $-80°C$ and then subsequently warmed to room temperature, always spontaneously decomposes to liquid SO_3. Although samples of α-SO_3 at first have very low vapour pressures, the vapour pressure always increases in time and liquid SO_3 begins to form. Over a period of several months, conversion of α- to γ-SO_3 is complete. An irreversible, exothermic transition has been observed at $-65°C$ when α-SO_3 is warmed from $-80°$. Probably monomeric SO_3 molecules are present in the low-temperature condensate and these react to give a mixture of the cyclic polymer, S_3O_9, and long-chain polymers, α-SO_3. In time, α-SO_3 changes slowly to S_3O_9.

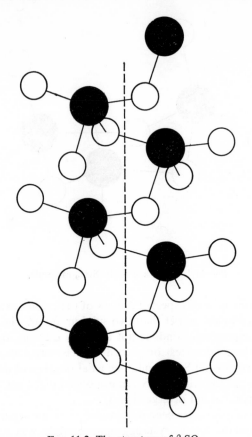

FIG. 11.2. The structure of β-SO$_3$.

β-SO$_3$ crystals are only formed if small quantities of water are present in SO$_3$ samples, for example, a number of H$_2$O/SO$_3$ mixtures, in which the mole percentage of SO$_3$ varies from 80 to 99·9, all solidify to crystals of β-SO$_3$. β-SO$_3$ is believed to consist of polysulphuric acids of general formula, HO(SO$_3$)$_n$SO$_2$OH. The function of the water necessary for the formation of this modification is to terminate chains on linked SO$_3$ molecules. Although it has been possible to locate the positions of sulphur and oxygen atoms by X-ray diffraction, this technique can, unfortunately, tell us nothing about the nature of the terminating groups.

Little is known about the gelatinous form of SO$_3$. This can be made by warming samples in an early stage of polymerization to β-SO$_3$ and it has been suggested this may represent an early stage of cross-linking.[13]

Stabilization of Liquid Sulphur Trioxide

It is often convenient to transport, store and use sulphur trioxide in the liquid state. The temperature range over which this exists is fairly narrow (b.p. = 44·5°C) and polymerization readily occurs. If traces of β-SO$_3$ are present on freezing, polymerization in the solid state will continue because γ-SO$_3$ has a higher vapour pressure than β-SO$_3$ and is slowly converted into this form by the diffusion of SO$_3$ molecules.

Polymerization of the liquid can be prevented, even when water is present, if the temperature is maintained at about 30 °C. This can conveniently be done for short periods of time but not for long storage or for transportation. Consequently, the possibility of stabilizing the liquid by the addition of substances which inhibit polymerization has been investigated.[13] Certain boron compounds are among the most effective inhibitors. Those which have been used include B_2O_3, BCl_3, BF_3, $Na_2B_4O_7$, KBF_4 and $NaBF_4$. Only small quantities, not more than 1·5% by weight of boron, are needed. The additive is mixed with SO_3 and the mixture heated for some hours to 60–100 °C. Organo-boron compounds like methyl borate and the dimethyl ether complex of BF_3 also act as inhibitors and with these, no heat treatment is needed. A wide variety of organic additives has been recommended. These include esters, amides and chloro-derivatives of sulphonic acids, aromatic amines, amides and nitrocompounds, benzoic acid and dimethyl sulphate.

It has been proposed[14] that polymerization proceeds in two stages:

(a) The formation of polysulphuric acids in the liquid phase according to

$$H_2S_{n-1}O_{3n-2} + SO_3 \rightleftharpoons H_2S_nO_{3n+1}$$

(b) Nucleation of a solid phase,

$$xH_2S_nO_{3n+1} \text{ (liquid)} \rightarrow xH_2S_nO_{3n+1} \text{ (solid)}.$$

Inhibition of the growth of solid polyacids requires the interaction of the additive with the terminal hydroxyl groups. In some cases, this may be a chemical reaction between the additive and the small amounts of polyacids initially present to convert them to complexes which cannot react with further sulphur trioxide. Boron compounds may well function in this way in view of the probable formation of complexes of the kind $H[B(HSO_4)_4]$ and $B(HS_2O_7)_3$ in solutions of boron compounds in oleum. With many other additives, where a specific chemical reaction is unlikely, it is probable that hydrogen bonding between the terminal hydroxyl groups of the polyacids and the polar additive molecules prevents solid nucleation.

Structure in the Gaseous and Liquid States

The electron diffraction study[15] of the vapour showed the molecule to be planar with S,O bond length = 0·143 ± 0·002 nm and the O–S–O angle = 120 ≦ 2°. The O,O distance = 0·248 ± 0·003 nm. Planarity is confirmed by measurements of the dielectric constant of the vapour at different temperatures which showed the molecule has zero dipole moment.

Much of our knowledge about sulphur trioxide in the liquid state comes from detailed studies of the vibrational spectra and comparison with spectra of gas samples. The planar SO_3 molecule belongs to the symmetry group, D_{3h}, and possesses six vibrational degrees of freedom constituting the following four fundamental frequencies: A_1' (R, p); A_2'' (i.r.); and two E' (R, dp, i.r.) where R, i.r., p and dp stand for Raman active, infrared active, polarized and depolarized respectively. The four fundamental modes are shown in Fig. 11.3. ν_1 is the totally symmetric stretching vibration, ν_2 is an out-of-plane vibration, ν_3 and ν_4 are the asymmetric stretching and deformation vibrations. ν_1 and ν_2 correspond with non-degenerate vibrations while ν_3 and ν_4 correspond with doubly-degenerate vibrations (E').

The vibrational spectrum has proved difficult to measure because of the extreme chemical reactivity of sulphur trioxide towards the halide salts conventionally used as window

107

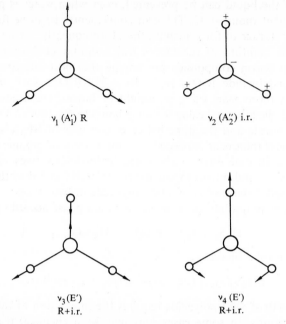

$\nu_1 (A_1')$ R $\nu_2 (A_2'')$ i.r.

$\nu_3 (E')$ $\nu_4 (E')$
R+i.r.   R+i.r.

FIG. 11.3. Fundamental vibrations of the planar SO_3 molecule.

materials in infrared spectrometers. It has recently[16] been measured on gaseous samples in cells constructed of Pyrex tubing to which thin windows of silicon were fused and also in the form of a xenon matrix at liquid nitrogen temperature. In the latter case, halide windows can be used without being attacked by the trioxide. The bands observed and their assignments are given in Table I. The three infrared active bands are present in both spectra. The gaseous sample shows, in addition, two weak bands, one an overtone and one a combination band. Raman lines have been observed[17] in gaseous SO_3 at 531·5, 1068 and 1389 cm^{-1} and identified respectively with ν_4, ν_1, and ν_3. The infrared results have been confirmed subsequently by measurements on gaseous mixtures of sulphur tri-

TABLE I

Assignments of the Vibrational Spectra of Monomeric Sulphur
Trioxide in the Gas Phase and, at Liquid Nitrogen Temperature,
in a Xenon Matrix
ν_1 is the Raman displacement observed at 1068 cm^{-1}

Absorption frequencies (cm^{-1})			
Gas phase		Xenon matrix	
Ref. 18	Ref. 16	Ref. 16	Assignment
2777	2773 vw		$2\nu_3$
2454	2443 w		$\nu_1 + \nu_3$
1391	1391 vs	1404 / 1373 vs	ν_3
529	529 s	525 m	ν_4
496	495 ms	464 s	ν_2

oxide and nitrogen circulated slowly through the spectrometer cell at room temperature. Cell windows were made of a polymeric material, "Irtran 2", for measurements between 650 and 4,000 cm⁻¹ and of potassium bromide coated with a very thin film of fluorocarbon grease for use below 650 cm⁻¹.

The Raman spectrum of liquid SO_3 is much more complex than that of the vapour. In addition to the three lines which characterize the SO_3 molecule, many other lines are observed (Fig. 11.4).[19] These are attributable to a polymeric form of the compound. At room temperature, Gillespie and Robinson observed 25 lines, 5 of which were polarized, in liquid sulphur trioxide obtained by distillation from 65% oleum. It appears that the polymer of prime importance is the non-planar cyclic trimeric molecule, S_3O_9, and the liquid consists mainly of a mixture of this with the monomer. When the temperature is raised, many of the Raman lines disappear and at 100 °C the spectrum shows only those lines characteristic of the monomer indicating depolymerization on heating.

The identification of the polymer is based on two pieces of evidence.

Firstly, the number of fundamental frequencies observed in the Raman spectrum is consistent more with S_3O_9 rings in the 'chair' configuration than with any alternative. There are twenty normal vibrations of the S_3O_9 ring in this configuration (Fig. 11.5 and Table II). The three A_2 modes are inactive in the Raman and infrared spectra. The other seventeen are made up of seven ring vibrations and ten involving the SO_2 groups and should all occur in both spectra. Assignments of the Raman spectrum have been based on the 'chair' configuration, which has C_{3v} symmetry. All except a few of the lines of least intensity can be accounted for in this way. These were attributed by Gillespie and Robinson to the presence of another polymeric species in small amounts, the most likely being the 'boat' form of the trimer.

Secondly, the Raman lines which characterize the solid γ-form are all found to occur quite strongly in the spectrum of the liquid and it is reasonable to suggest the persistence of S_3O_9 rings in the liquid state.

The infrared spectrum of pure liquid sulphur trioxide was reported[20] for the first time in 1966. Even small traces of water produce gross changes in the spectrum and it is ne-

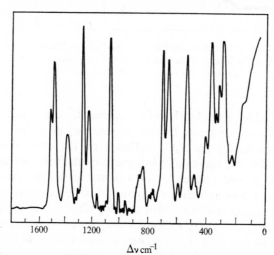

$\Delta v\,cm^{-1}$

FIG. 11.4. The Raman spectrum of liquid sulphur trioxide (reproduced by permission of the National Research Council of Canada from the *Canadian Journal of Chemistry*, **39**, 2191–8 (1961)).

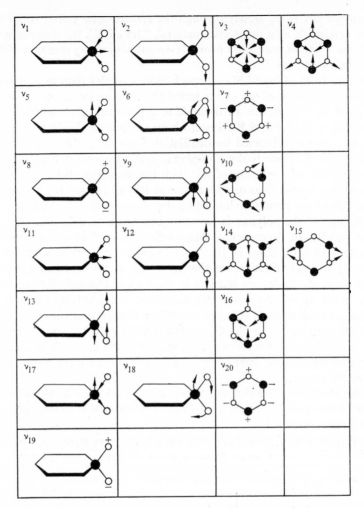

FIG. 11.5. The fundamental vibrations of the S_3O_9 molecule (reproduced by permission of the National Research Council of Canada from the *Canadian Journal of Chemistry*, **39**, 2191—8 (1961)).

cessary to dry the trioxide scrupulously before examination. For this reason, earlier infra-red spectroscopic data on the liquid cannot be regarded as entirely reliable. The assign-ments made by Stopperka for the S_3O_9 molecule on the basis of infrared and Raman spectra are given in Table III. These differ in some respects from Gillespie and Robinson's[19] and it is advisable to regard neither set of assignments as definitive. Indeed, the high-resolution Raman spectra of anhydrous sulphur trioxide measured over the temperature range 18–85°C shows some features which are not readily explicable on the basis of an equilibrium between SO_3 and S_3O_9 molecules only.[21] The bands which characterize the monomer (those at 1068 and 1389 cm⁻¹) become more intense with increase in temperature. All the others except one become less intense with temperature increase and these are due to the trimer. The exception is the band observed at 532 cm⁻¹, with its shoulder at 547 cm⁻¹, which remains constant in intensity and shape throughout this temperature range. Gillespie

TABLE II

Normal Vibrations of the S_3O_g Molecule

In the A_1 and A_2 vibrations the three SO_2 groups move in phase. In the
E vibrations, they move out of phase.

Class	Activity	Description	Vibration
A_1	i.r. + R(p)	SO_2, sym. stretch	ν_1
		SO_2, in-phase bending	ν_2
		Ring, sym. contraction	ν_3
		Ring, sym. deformation	ν_4
		SO_2, asym. stretch	ν_5
		SO_2, rocking	ν_6
		Ring, sym. deformation	ν_7
A_2	Inactive in both	SO_2, torsional	ν_8
		SO_2, rocking	ν_9
		Ring, vibrations	ν_{10}
E	i.r. + R	SO_2, sym. stretch	ν_{11}
		SO_2, bending	ν_{12}
		SO_2, rocking	ν_{13}
		Ring, vibrations,	ν_{14}
		Ring, vibrations,	ν_{15}
		Ring, vibrations,	ν_{16}
		SO_2, asym. stretch	ν_{17}
		SO_2, rocking	ν_{18}
		SO_2, torsional	ν_{19}
		Ring, asym. deformation	ν_{20}

TABLE III

Assignment of Fundamental Frequencies (cm^{-1}) in S_3O_9 according to Stopperka, K., *Zeit. anorg. all.
Chem.* **345**, 277 (1966).

Infrared	Raman	Vibration
1516	1516	ν_5
1492	1491	ν_{17}
1274	1272	ν_1
1232	1227	ν_{11}
865		ν_{15}
842	845	ν_{16}
772		ν_{14}
698	698	ν_3
666	665	ν_4
654	650	ν_6
588	598	ν_{18}
545		$\nu_2 + \nu_{12}$
412	408	ν_{13}
	369	ν_7
	317	ν_{19}
	289	ν_{20}

111

and Robinson attributed the band which they observed at 534 cm^{-1} in liquid SO_3 to the in-phase bending mode of the SO_2 group in the trimer and suggested it was coincident with the SO_2 bending mode (at 532 cm^{-1}) in the monomer. It is unlikely that such a mode would not be changed by polymerization although it is difficult to suggest any alternative explanation for the observations.

There is no doubt that the predominant polymeric species in liquid sulphur trioxide is the S_3O_9 molecule. Some weak absorption bands due to this have even been observed in the gas phase[16] but their intensities diminish rapidly with increase in temperature or decrease in pressure and the planar SO_3 molecule is the most important species in this phase.

References

1. WALRAFEN, G. E., *J. Chem. Phys.* **40**, 2326 (1964).
2. STOPPERKA, K., *Zeit. anorg. all. Chem.* **345**, 264 (1966).
3. GILBERT, E. E., *Chem. Rev.* **62**, 549 (1962).
4. SMITS, A., and SCHOENMAKER, P., *J. Chem. Soc.* 2554 (1924).
5. SMITS, A., and SCHOENMAKER, P., *J. Chem. Soc.* 1108 (1926).
6. SMITS, A., and SCHOENMAKER, P., *J. Chem. Soc.* 1603 (1926).
7. GERDING, H., and MOERMAN, N. F., *Z. Physik. Chem.* B **35**, 216 (1937).
8. WESTRIK, R., and MacGILLAVRY, C. H., *Rec. Trav. Chim.* **60**, 794 (1941).
9. McDONALD, W. S., and CRUICKSHANK, D. W. J., *Acta Cryst.* **22**, 48 (1967).
10. PASCARD, R., and PASCARD-BILLY, C., *Acta Cryst.* **18**, 830 (1965).
11. CRUICKSHANK, D. W. J., *J. Chem. Soc.* 5486 (1961).
12. WESTRIK, R., and MacGILLAVRY, C. H., *Acta Cryst.* **7**, 764 (1954).
13. BEVINGTON, C. F. B., and PEGLER, J. L., *Chemical Society Special Publication* No. 12, 283 (1958).
14. ABERCROMBY, D. C., HYNE, R. A., and TILEY, P. F., *J. Chem. Soc.* 5832 (1963).
15. PALMER, K. J., *J. Amer. Chem. Soc.* **60**, 2360 (1938).
16. LOVEJOY, R. W., COLWELL, J. H., EGGERS, D. F., Jr., and HALSEY, G. D., Jr., *J. Chem. Phys.* **36**, 612 (1962).
17. GERDING, H., NIJVELD, W. J., and MULLER, G. J., *Z. Physik. Chem.* B **35**, 193 (1937).
18. BENT, R., and LADNER, W. R., *Spectrochim. Acta* **19**, 931 (1963).
19. GILLESPIE, R. J., and ROBINSON, E. A., *Can. J. Chem.* **39**, 2189 (1961).
20. STOPPERKA, K., *Zeit. anorg. all. Chem.* **345**, 277 (1966).
21. VANDORPE, B., and MIGEON, M., *Rev. Chim. Min.* **2**, 303 (1965).

CHAPTER 12

TETRASULPHUR TETRANITRIDE

Introduction

Several compounds are known which contain only sulphur and nitrogen. Tetrasulphur tetranitride, S_4N_4, first made in 1835 by the reaction between disulphur dichloride, S_2Cl_2, and ammonia,[1,2] is the most stable of these. Other compounds of general formula $(SN)_n$ are known in which $n = 1, 2$ and ∞. The SN radical is known only as a short-lived species in the vapour state. S_2N_2 is stable at low temperatures but explodes violently above ambient temperature. S_4N_4 is stable at temperatures up to its melting-point (187°C) but is liable to explosive decomposition on heating or on mechanical shock. These compounds form a unique series and their interrelationship, particularly the nature of the S,N bonds, is of great interest. Severe limitations are imposed on the experimental study of SN and S_2N_2 by their instability and very much more is known of the properties, structure and reactions of S_4N_4.

The chemistry of S_4N_4 is distinguished by its molecular structure, a puckered ring of alternate S and N atoms, by the strength of the S,N bonds and by its ability to act as a Lewis base or acid. Its reactions have been extensively studied and it is the starting material for the synthesis of many novel compounds, such as those containing S–N-metal ring systems.

There is experimental evidence to suggest that delocalization of π-electrons occurs around the ring and several theoretical treatments of the electronic structure of S_4N_4 have been published.

Preparation of S_4N_4 and Other S,N Compounds

The reaction between S_2Cl_2 and NH_3 has been examined in various experimental conditions. The yield of S_4N_4 is good if S_2Cl_2 is dissolved in carbon tetrachloride and chlorinated to form SCl_2 prior to the addition of ammonia.[3]

$$6S_2Cl_2 + 16NH_3 \rightarrow S_4N_4 + 8S + 12NH_4Cl$$
$$6SCl_2 + 16NH_3 \rightarrow S_4N_4 + 2S + 14NH_4Cl$$

Ammonia is passed through the solution for 2 hr during which time S_4N_4 separates out as an orange–yellow solid. The temperature of the reaction mixture must be controlled

113

below 50°C. Impure S_4N_4 is recrystallized from dioxane and benzene and repeated purifications in this way give samples which melt at 187°C. The tendency of S_4N_4 to decompose explosively appears to increase with the purity of the sample.[3]

Another preparative method reported recently[4] is the reaction at −95°C between NH_3 and SF_4. This is a highly exothermic reaction and produces S_4N_4 in yields up to 70%.

$$12SF_4 + 64NH_3 \rightarrow 3S_4N_4 + 2N_2 + 48NH_4F$$

S_4N_4 can be extracted with benzene from the reaction product, leaving a residue of insoluble material, believed to be $(SN)_x$ polymer.

An unusual ring expansion reaction can also be used to prepare S_4N_4. The thiotrithiazyl cation, $S_4N_3{}^+$, is a planar seven-membered ring. When S_4N_3Cl is added to a solution of aluminium azide, $Al(N_3)_3$, in tetrahydrofuran, nitrogen is released and S_4N_4 can be recovered in 85–90% yield.[5]

$$3S_4N_3Cl + Al(N_3)_3 \rightarrow 3S_4N_4 + 3N_2 + AlCl_3$$

Disulphur dinitride, S_2N_2, can be made by the thermal decomposition[6] of S_4N_4 in the presence of silver under a vacuum. The presence of silver prevents the combination of S_2N_2 with the free sulphur also formed in the reaction, by removing it as Ag_2S. Polymeric $(SN)_x$ is also formed at the same time and can be collected by condensation on a water-cooled cold finger. S_2N_2 is condensed on a cold finger cooled in liquid nitrogen. It is stable for extended periods at −80°C but turns black eventually due to the formation of $(SN)_x$. S_2N_2 is a white solid which explodes at temperatures above 30°C to give sulphur and nitrogen.

$(SN)_x$ forms fibre-like crystals and is believed to be a linear polymer. It is a diamagnetic compound which possesses semiconductor properties, its electrical resistence decreasing with increasing temperature.

Reactions

S_4N_4 is regarded as a nitride rather than a sulphide because ammonia and not hydrogen sulphide is formed on hydrolysis. Sulphur oxo-acids are also formed, the particular ones depending on the conditions for hydrolysis:
In dilute alkali:

$$2S_4N_4 + 6OH^- + 9H_2O \rightarrow 2S_3O_6{}^{2-} + S_2O_3{}^{2-} + 8NH_3$$

In concentrated alkali:

$$S_4N_4 + 6OH^- + 3H_2O \rightarrow 2SO_3{}^{2-} + S_2O_3{}^{2-} + 4NH_3$$

The hydrolysis products are typical for a compound containing sulphur in the +3 oxidation state.

Reduction with tin(II) or dithionite converts sulphur from +3 to +2 and tetrasulphur tetraimide, $S_4N_4H_4$, is formed. This is a white solid which is readily oxidized back to S_4N_4 by chlorine. The infrared spectrum of $S_4N_4H_4$ shows strong bands at 3220, 3285 and 3320 cm^{-1} which are typical of N–H bonds. There are no bands around 2600 cm^{-1} where S–H bonds would absorb and so the reduction product must be formulated as $S_4(NH)_4$. Complete breakdown of S_4N_4 by further reduction gives ammonia but no hydrazine.

$$S_4N_4 + 20HI \rightarrow 4H_2S + 4NH_3 + 10I_2$$

It may therefore be concluded that N–N bonds are absent in S_4N_4.

Treatment of S_4N_4 dissolved in CCl_4 with AgF_2 fluorinates the ring to give $(SF)_4N_4$. If the reaction mixture is heated, gaseous SN_2F_2 is formed. This, in turn, decomposes at about 250°C to NSF, thiazyl fluoride.

The chlorination of S_4N_4 yields $S_3Cl_3N_3$, trithiazylchloride. This molecule has a six-membered ring structure and is presumably formed by the breakdown of the S_4N_4 ring into smaller species, for example SN radicals, which then combine with the halogen. The product becomes stabilized by polymerization to the cyclic molecule.

$S_3Cl_3N_3$ reacts with $S_4(NH)_4$ to regenerate S_4N_4.

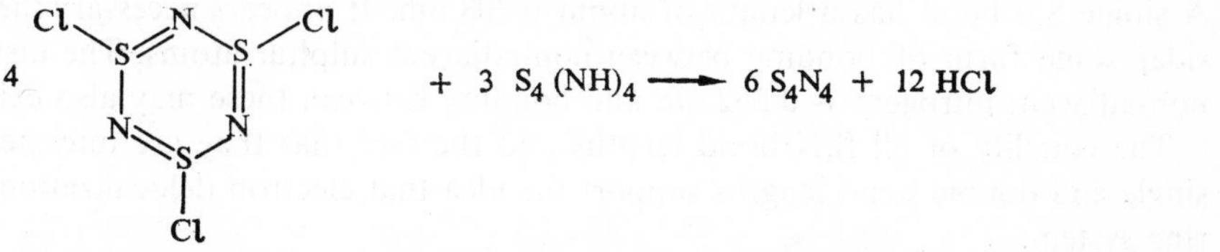

$$4 \quad [\text{structure}] \quad + \ 3 \ S_4(NH)_4 \longrightarrow 6 \ S_4N_4 + 12 \ HCl$$

If the HCl is not removed by the addition of a base, the adduct $S_4N_4 \cdot HCl$ is formed. This decomposes to give thiotrithiazylchloride, S_4N_3Cl, and NH_4Cl and Cl_2.

Cleavage of the S_4N_4 ring system can be brought about by heating or by reaction with ammonia or with alcohols. The product of the reaction with NH_3 is the ammoniate of empirical formula, $S_2N_2 \cdot NH_3$. Even at ambient temperature, this breaks down into S_2N_2 and NH_3. With methanol, S_4N_4 reacts to give an analogous adduct, $S_2N_2 \cdot CH_3OH$.

Complexes have been prepared between S_4N_4 and many metal halides.[7] For example, 1:1 adducts between S_4N_4 and $TiCl_4$, VCl_4, WCl_4, $SbCl_5$, BF_3 and $MoCl_5$, can be made by refluxing the halide with S_4N_4 in CCl_4 solution. The metal halide acts as a Lewis acid and the metal atom is coordinated to the S_4N_4 molecule through one nitrogen atom. This has been demonstrated by X-ray crystallography[8] which shows that the ring structure of S_4N_4 is preserved on coordination but that its shape is altered considerably (see below).

In polar solvents, S_4N_4 reacts with metal halides to give a variety of products including some where the metal atom, as well as sulphur and nitrogen, forms part of a ring system.[9] For example, $Pd(NS_3)_2$ consists of two five-membered rings linked by the palladium atom. Nickel carbonyl reacts with S_4N_4 in benzene solution to form $Ni(S_4N_4)$. This is a black compound which is insoluble in all common solvents but treatment with a solvent such as alcohol or acetone very slowly converts some of it to $Ni(HN_2S_2)_2$ in solution. The nickel atom is chelated by two S_2N_2 groups:

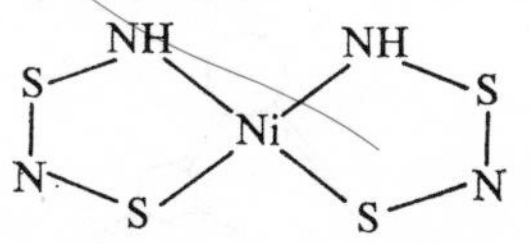

Many other complexes have been made between S_4N_4 and metals of Group VIII. For example, the Co, Pd and Pt analogues of $Ni(HN_2S_2)_2$ are known. They are typical chelate complexes, insoluble in water but soluble in most organic solvents, highly coloured and very stable.

Structure of S_4N_4

S_4N_4 forms monoclinic crystals belonging to the space group, $P2_1/n(C^5_{2h})$. Buerger[10] has determined the crystalline structure by X-ray diffraction. The unit cell dimensions are: $a = 0.874$, $b = 0.714$ and $c = 0.8645$ nm; $\beta = 92°21$ and z, the number of molecules per unit cell, $= 4$.

In 1952, Clark[11] verified the unit cell and space group reported by Buerger. In each S_4N_4 molecule, the four sulphurs form a slightly distorted tetrahedron and the four nitrogens all lie in one plane and, within experimental error, are situated at the corners of a square. All S,N bond lengths are the same (0·162 nm). This value is intermediate between the values expected for single and double S,N bonds (0·174 and 0·154 nm respectively). Pairs of sulphur atoms which are not linked through the same nitrogen atom are only 0·258 nm apart. The van der Waals radius of sulphur is 0·185 nm and so the distance of closest approach of two sulphur atoms not chemically bound together should be 0·37 nm. A single S,S bond has a length of about 0·208 nm. It appears necessary therefore to consider some form of bonding between non-adjacent sulphur atoms. The distance between non-adjacent nitrogens is 0·252 nm and bonding between these may also exist.

The equality of all S,N bond lengths and the fact that they are intermediate between single and double bond lengths support the idea that electron delocalization occurs in the ring system.

On the basis of electron diffraction results, Lu and Donohue[12] proposed a cyclic structure of D_{2d} symmetry (I) (Fig. 12.1). This structure is closely similar to that of realgar, As_4S_4. Their data confirm all the S,N bond lengths are equal (S,N = 0·162 nm), S,S = 0·269 nm and the bond angles are SNS = 112° and NSN = 106°.

An alternative structure, the cage molecule (II), has been proposed from another X-ray diffraction study.[13] (II) also has D_{2d} symmetry but differs from (I) in that the nitrogen

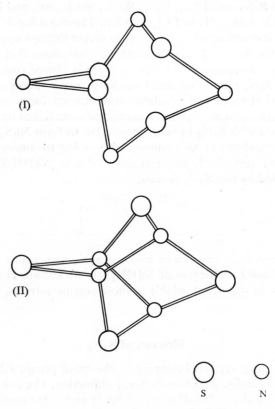

FIG. 12.1. Proposed structures for S_4N_4.

atoms of the eight-membered ring are bonded together in pairs. At first sight, (II) is more attractive than (I) because the S and N atoms form two and three bonds respectively, the numbers to be expected from the group valence of the two elements. Support for (II) was also adduced from measurements of the infrared and Raman spectra of S_4N_4.[14] In particular, lines at 888 and 934 cm^{-1} in the Raman spectrum were identified respectively with symmetric and asymmetric stretching frequencies of the N,N bonds. In a more recent analysis of the vibrational spectrum,[15] discussed in detail below, these bands have been assigned to S,N stretching frequencies and there is no unequivocal spectroscopic evidence to favour structure (II).

The X-ray diffraction study by Sharma and Donohue[16] has confirmed the conclusions of Clark and Lu and Donohue and structure (I) is now generally accepted as correct. The average S,N bond length = 0·1616 ± 0·001 nm. The geometry of the molecule is illustrated in Fig. 12.2.

The classification of S_4N_4 as D_{2d} results from the various symmetry elements which (I) possesses. To simplify the pictorial representation of the molecule, the sulphur atoms are assumed to lie at the apices of a regular tetrahedron. This is shown in Fig. 12.3 where the sulphurs are located at the opposite corners of a cube, two above and two below the plane

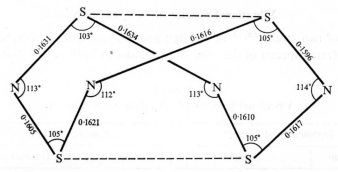

FIG. 12.2. The structure of the S_4N_4 molecule as determined by X-ray diffraction (data from Sharma, B. D., and Donohue, J., *Acta Cryst.* **16**, 891 (1963)).

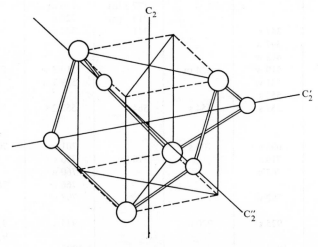

FIG. 12.3. The symmetry elements of S_4N_4.

of nitrogens. The four nitrogens are situated outside the cube on the normals through the centres of the four vertical faces. There are three mutually perpendicular two-fold axes, C_2, C_2' and C_2''. There are also two vertical dihedral planes σ_v and σ_v' bisecting the angles between C_2' and C_2''. Hence the molecule belongs to the symmetry group D_{2d}.

If the positions of the sulphur and nitrogen atoms in (I) are interchanged so that the sulphurs are co-planar and the nitrogens are tetrahedrally arranged, the symmetry of the molecule remains the same.

Significant changes take place in the shape and dimensions of the S_4N_4 molecule when coordinated to metal halides. The X-ray diffraction study of $S_4N_4BF_3$[17] has shown that the BF_3 molecule is bonded via boron to one nitrogen of the S_4N_4 ring. The four sulphurs are situated roughly at the corners of a square and the nitrogen atoms are alternately above and below this. The S,N bond lengths vary between 0·1666 and 0·1542 nm. The configuration of the ring is similar in $S_4N_4 \cdot SbCl_5$[8] and evidently very different from that of the S_4N_4 molecule itself. In both adducts, the S,S distance is greater than 0·38 nm. The formation of an adduct has clearly destroyed whatever bonding may exist between pairs of sulphur atoms in S_4N_4.

Vibrational Spectra of S_4N_4 and S_2N_2

The results of two studies[14,15] of infrared and Raman spectra of S_4N_4 are given in Table I. The infrared spectra of the solid and solutions in benzene and dioxane have been

TABLE I
Vibrational Spectrum of S_4N_4 (frequencies in cm^{-1})

Lippincott and Tobin[14]			Bragin and Evans[15]		
Raman	Infrared		Raman	Infrared	
Dioxane solution	Solid	Solution	Dioxane solution	Solid	Solution
177			177·5 (d)	190 m	
213			213 (p)	226 w	
347	347 s		341	347 s	330 m
	397 w				
	412 w				
519	519 w	524 m		519·3 w	519·3 w
	531 w			529·7 w	
	552 s				
561	557 s	554 s	564	552·2 s	556·5 m
615					
					702 m
	696 s	700 s		701 vs	705 vs
720	719 s		716 (p)	726 s	
	762 w			760 w	
785				766 w	766 vw
	792 w			798 w	
888					
934	925 s	920 m		925 s	938 s
					938·5 m
	1000 w			1007 w	

measured. The Raman spectrum cannot be obtained on the melt because of its explosive properties and has been measured only on saturated solutions in benzene and dioxane.

The S_4N_4 ring with D_{2d} symmetry has fourteen fundamental vibrations. The symmetry species are $3A_1 + 2A_2 + 2B_1 + 3B_2 + 4E$. Two of these, $2A_2$, are inactive in both the Raman and infrared spectra. The remaining twelve are all Raman active and seven of these ($3B_2 + 4E$) are active in the infrared as well. Vibrational bands observed have been assigned on the basis of an internal coordinate set consisting of S–N and S–S stretching modes and SSN bending modes (Table II). It is interesting to note that structures (I) and (II) both have D_{2d} symmetry and so the vibrational spectrum does not provide a means of distinguishing between them. In fact, assignments have been made on the basis of either structure[14,15] but, in view of the evidence from other sources that (I) is the correct structure of S_4N_4, the assignment using this model is to be preferred.

TABLE II

Assignment of Vibrational Frequencies for S_4N_4

Class	Assignment	Frequency (cm^{-1})
A_1	S–N stretch	716
	SSN bend	529·7
	S–S stretch	213
A_2	S–N stretch	557*
	SSN bend	428*
B_1	S–N stretch	888
	SSN bend	615
B_2	S–N stretch	705
	SSN bend	564
	S–S stretch	177·5
E	S–N stretch	938
	S–N stretch	766
	SSN bend	519·3
	SSN bend	341

* Calculated values.

The structure of S_4N_4 leads one to expect a great deal of mixing of the bending and stretching vibrations. Thus each SSN angle shares its central atom with a S,N bond which lies in the plane of the angle. This bond is normal to the S,S bond and an extension of the angle displaces the shared sulphur atom and stretches the S,N bond. Thus an SSN bending motion must involve S,N bond stretching. This factor, together with the internal co-ordinate mixing of the normal vibrational modes which can also occur, brings together the stretching and bending frequencies so that there is comparatively little difference between them.

The infrared spectrum of S_2N_2 has been interpreted in terms of a planar ring of alternate S and N atoms.[15,18] This has D_{2h} symmetry. This is an important application of infrared spectroscopy because the marked instability of the compound makes its examination difficult by other physical methods. The interconversion of S_2N_2 and S_4N_4 supports the view that these molecules contain similar bonds. The ring structure of S_2N_2 distinguishes it from most other tetratomic molecules which are either linear or planar (e.g. C_2N_2, N_2O_2 and N_2F_2).

Strength of Sulphur–Nitrogen Bonds

An estimate of this can be obtained from the heat of formation of S_4N_4 and other thermochemical data. The heat of formation has been determined by studying the decomposition of S_4N_4 on a heated platinum wire to form the elements, amorphous sulphur and nitrogen gas. The value found was 460 ± 8.4 kJ mol^{-1}. From this and the standard heats of formation of $S(g) = 274.5$ kJ mol^{-1} and $N(g) = 473.0$ kJ mol^{-1} and the heat of sublimation of S_4N_4 (estimated value $= 62.7 \pm 20.9$ kJ mol^{-1}), the average S–N bond strength in S_4N_4, $D(S-N)_{av} = 308$ kJ mol^{-1}. This calculation is based on the thermochemical cycle shown and it is assumed that only S,N bonds need be considered in S_4N_4.

$$4S(g) + 4N(g) \xrightarrow{8D(S-N)av} S_4N_4(g)$$
$$\uparrow -1098 \quad \uparrow -1892 \quad \uparrow -62.7$$
$$4S(s) + 2N_2(g) \xrightarrow[-460.0]{} S_4N_4(s)$$

Heat evolved is positive : heat absorbed is negative.

The value of $D(S-N)_{av}$ may be compared with the value of $D(S-N) = 482 \pm 105$ kJ mol^{-1} in NS and NS$^+$ obtained from a partial analysis of the band spectra of these species in the gas phase. These data are consistent with relative bond orders of 2.5 in NS and 1.65 in S_4N_4.[20]

It appears that S,S or N,N interactions only make small contributions, probably less than 84 kJ mol^{-1}, to the total bonding energy in S_4N_4.

Dipole Moment

Measurements of the dipole moment in benzene and carbon disulphide have shown[21] apparent moments of 0.52 and 0.72D respectively. These cannot easily be reconciled with the symmetrical structure of the S_4N_4 molecule. The moments have been attributed to abnormally large atomic polarization in this molecule which could be only slightly polar or even non-polar.

Bonding

In deriving a theoretical description of the bonding within S_4N_4, there are several properties of the molecule which must be borne in mind:
1. The equality of all S,N bond lengths.
2. The shortness of the distance between pairs of sulphur atoms not linked through the same nitrogen.
3. The preference for a structure in which the nitrogen atoms are co-planar and the sulphurs are tetrahedrally arranged rather than the converse.

The extent to which the d-orbitals of sulphur participate in bonding must also be considered. Finally, it should be possible to explain how several binary sulphur nitrides can exist and to describe the bonding in them.

According to resonance theory, the equality of S,N bonds results from resonance between structures such as:

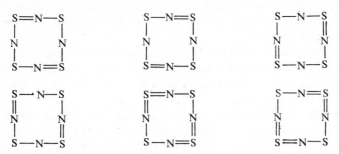

This description accounts for the intermediate value of the S,N bond length between those of single and double S,N bonds. In view of (2) there should also be included a contribution to the resonance hybrid from cross-bonded structures like:

From the observed S,N bond lengths in various compounds, Chapman and Waddington[20] have calculated that the bond order in S_4N_4 is 1·65. In their description of the bonding, there is a total of eight electrons in a delocalized system around the ring, each atom contributing one electron. The bond angles within the ring are accounted for on the assumption that the sulphur is approximately sp^3 hybridized, with one hybrid orbital occupied by a lone pair, and the nitrogen is approximately sp^2 hybridized with one hybrid orbital also occupied by a lone pair. Each sulphur has one electron in a d orbital and each nitrogen has one electron in a p-orbital and these are involved in d_π–p_π bonding to form the delocalized system.

These authors propose that the delocalized electrons can be regarded as restricted to but free to move on a spherical surface, so defined that all eight atoms lie on or very close to it. Just as in isolated atoms, the electrons on this surface can have different energies. They interpreted the ultraviolet absorption spectrum of S_4N_4 in terms of transitions between the various possible energy levels. Solution of the appropriate wave-equation shows these levels are defined by:

$$E_l = \frac{h^2 l (l + 1)}{8\pi^2 \mu a^2}$$

where a = radius of the sphere, μ = reduced mass of the molecule, and l is a quantum number which can have values 0, 1, 2, etc., and m is another quantum number which has values of $-l, \ldots, 0, \ldots, +l$, leading to degeneracy of each energy level for which $l > 0$. The electronic levels in order of increasing energy are:

$$6 \; -\;-\;-\;-\;- \quad l = 2; \quad m = -2, -1, 0, +1, +2$$

E
(in units of ↑ $\quad 2 \;\; \uparrow\downarrow \;\; \uparrow\downarrow \;\; \uparrow\downarrow \quad l = 1; \quad m = -1, 0, +1$
$h^2/8\pi^2\mu a^2$) $\qquad\quad 0 \qquad\;\; \uparrow\downarrow \qquad\quad l = 0; m = 0$

For the neutral molecule the eight π-electrons fill the energy levels where $l = 0$ and 1. They form a quasi s^2p^6 closed shell and the molecule is diamagnetic. Taking the radius of the sphere as 0·178 nm, the wavelength for the main electronic transition in the ultraviolet region assumed to be of the type was calculated as 256 nm, in good agreement with the experimental value of 257 nm.

The electron-on-a-sphere model has been criticized[22] for not explaining the low energy bands observed below 257 nm in the spectrum of S_4N_4. The transition at 257 nm has been alternatively assigned to a $\pi \rightarrow \pi^*$ transition in the weak S,S bond, which results in an opening-out of the cage structure.

The extended Hückel M.O. theory has been applied[23] by Turner and Mortimer to the bonding in S_4N_4. It was concluded that the d-orbitals on the sulphur may contribute to but do not play a major part in describing the electronic structure. The inclusion of these orbitals in the calculations has little effect on the energies of the molecular orbitals. The structure involving co-planar nitrogens is energetically favoured over the alternative with co-planar sulphurs by about 134 kJ mol^{-1}. Significant bonding is expected between sulphurs located on the same side of the plane containing the four nitrogens but no bonding is expected between pairs of nitrogens.

The various M.O. descriptions of S_4N_4 can account for the observation[24] that paramagnetic anions are formed from S_4N_4 by treatment of it with potassium in dimethoxyethane solution. A green solution is obtained which shows an E.S.R. spectrum consisting of nine lines. This is consistent with the presence of delocalized electrons in a paramagnetic ion such as $S_4N_4^-$. More highly-charged anions, up to $S_4N_4^{4-}$, are also produced in this reaction. $S_4N_4^{4-}$ itself is diamagnetic. It is the conjugate base of $S_4(NH)_4$ which has a structure like that of the S_8 molecule. Presumably the diamagnetism results from a collapse of the cage-like structure on addition of the fourth electron.

A detailed critique of the several studies of the bonding in S_4N_4 together with the latest experimental N.M.R. data has recently been published.[25] The ^{14}N N.M.R. spectrum for the molecule is found to consist of a single line at $+485 \pm 20$ ppm from the resonance of a saturated aqueous solution of a nitrite. This agrees with the equivalence of the nitrogen nuclei but it occurs at an unexpectedly high field. The chemical shift is much nearer to the shifts observed for single S,N bonds (530–540 ppm) than it is to the range observed for thiazenes such as the $S_4N_3^+$ ion (200–300 ppm). ^{14}N resonances tend to move strongly downfield with double-bonding because of paramagnetic deshielding. In the case of S_4N_4, the high shielding has been explained in terms of the high symmetry of the S_4N_4 molecule compared with the flatter thiazenes.

The bonding within the polymer $(SN)_x$ has been discussed[26] in terms of delocalized electrons. Each nitrogen has a p-orbital containing one electron which contributes to a conjugated system and each sulphur has a p-orbital containing a lone pair of electrons available for delocalization. Each S,N unit thus contributes three π-electrons. The valence shell of the sulphur can also be expanded to involve d-orbitals so that some $p_\pi-d_\pi$ bonding may occur. $(SN)_x$ shows semiconductor properties which are consistent with electron delocalization. Its infrared spectrum shows evidence for the presence of two S,N bonds of different lengths so the extent of delocalization remains in some doubt.

References

1. GREGORY, M., *J. Pharm.* **21**, 315 (1835).
2. GREGORY, M., *J. Pharm.* **22**, 301 (1835).
3. *Inorganic Syntheses*, vol. IX, p. 98.

4. COHEN, B., HOOPER, T. R., and PEACOCK, R. D., *J. Inorg. Nucl. Chem.* **28**, 919 (1966).
5. BECKE-GOERING, M., and MAGIN, G., *Z. Naturforsch.* **20**B, 493 (1965).
6. BURT, E. P., *J. Chem. Soc.* **97**, 1171 (1910).
7. NEUBAUER, D., WEISS, J., and BECKE-GOERING, M., *Z. Naturforsch.* **14**B, 284 (1959).
8. NEUBAUER, D., and WEISS, J., *Z. anorg. all. Chem.* **303**, 28 (1960).
9. WEISS, J., *Fortschr. chem. Forsch.* **5**, 635 (1966).
10. BUERGER, M. J., *Amer. Min.* **21**, 515 (1936).
11. CLARK, D., *J. Chem. Soc.* 1615 (1952).
12. LU, C., and DONOHUE, J., *J. Amer. Chem. Soc.* **66**, 818 (1944).
13. HASSEL, O., and VIERVOLL, H., *Tidskr. Kemi. Berg. Met.* **3**, 7 (1943).
14. LIPPINCOTT, E. R., and TOBIN, M. C., *J. Amer. Chem. Soc.* **73**, 4990 (1951).
15. BRAGIN, J., and EVANS, M. V., *J. Chem. Phys.* **51**, 268 (1969).
16. SHARMA, B. D., and DONOHUE, J., *Acta Cryst.* **16**, 891 (1963).
17. DREW, M. G. B., TEMPLETON, D. H., and ZALKIN, A., *Inorg. Chem.* **6**, 1906 (1967).
18. WARN, J. R. W., and CHAPMAN, D., *Spectrochim. Acta* **22**, 1371 (1966).
19. BARKER, C. K., CORDES, A. W., and MARGRAVE, J. L., *J. Phys. Chem.* **69**, 334 (1965).
20. CHAPMAN, D., and WADDINGTON, T. C., *Trans. Faraday Soc.* **58**, 1679 (1962).
21. ROGERS, M. T., and GROSS, K. J., *J. Amer. Chem. Soc.* **74**, 5294 (1952).
22. BRATERMAN, P. S., *J. Chem. Soc.* 2297 (1965).
23. TURNER, A. G., and MORTIMER, F. S., *Inorg. Chem.* **5**, 906 (1966).
24. CHAPMAN, D., and MASSEY, A. G., *Trans. Faraday Soc.* **58**, 1291 (1962).
25. MASON, J., *J. Chem. Soc.* A, 1567 (1969).
26. CHAPMAN, D., WARN, R. J. W., FITZGERALD, A. G., and YOFFE, A. D., *Trans. Faraday Soc.* **60**, 294 (1964).

CHAPTER 13

PERCHLORIC ACID

Introduction

One fundamental property which characterizes the non-metallic elements is the formation of oxo-acids, compounds which contain hydrogen and oxygen and which dissociate in solution to give solvated protons and oxo-anions. Some oxo-acids are of great technological importance. These are nitric, sulphuric and phosphoric acids, which are produced annually in millions of tons for use in the chemical industry. Many other oxo-acids, although not used so extensively as these, are of particular value to both the practical and theoretical chemist. Perchloric acid, $HClO_4$, is an outstanding example. It has many uses in preparative and analytical chemistry and the study of its properties, reactions and structure has deepened our understanding of inorganic oxo-acids as a class of compounds.

Perchloric acid is regarded as a very strong acid because its dissociation in aqueous solution is very extensive even at high concentrations. The dissociation constant cannot be readily determined by conventional methods and new ones, based on Raman and nuclear magnetic resonance spectroscopy, have been developed to make this possible. The difficulty of comparing the strengths of strong acids in aqueous solution led Hammett to propose his acidity function. This describes the tendency of an acid like perchloric or sulphuric to transfer a hydrogen ion to an acceptor molecule over concentrations ranging from dilute aqueous solution to the pure acid.

Perchloric acid forms several hydrates. Another consequence of the great strength of the acid is that both its mono- and dihydrate are ionized in the crystalline state and contain the hydrated ions, H_3O^+ and $H_5O_2^+$, respectively. These hydrates provide excellent media for studying the structures of these ions, which play such a significant role in the properties of aqueous solutions, in a crystalline lattice.

Preparation

Perchloric acid was first made in 1816 by von Stadion,[1] who observed the formation of potassium perchlorate by the action of sulphuric acid on potassium chlorate. Distillation from this mixture gives an aqueous solution of perchloric acid. The monohydrate of perchloric acid was prepared by Sérullas[2] in 1831 and the anhydrous acid made for the first

124

time by Roscoe[2] in 1862 by distillation of potassium perchlorate with excess sulphuric acid. The vapour from this distillation condenses to a white crystalline mass, the monohydrate, which can itself be distilled to produce anhydrous perchloric acid and an azeotrope containing 72·5% perchloric acid which boils at 203°C. Distillation at atmospheric pressure causes some decomposition to chlorine, oxides of chlorine and oxygen.

The anhydrous acid has been made by vacuum distillation of the dihydrate and monohydrate: by the treatment of chlorine heptoxide (itself made by the dehydration of aqueous perchloric acid with excess phosphorus pentoxide) with the requisite amount of water,

$$Cl_2O_7 + H_2O \rightarrow 2HClO_4$$

and by the reaction of fuming sulphuric acid (15–20%) with 72% perchloric acid followed by vacuum distillation,

$$HClO_4 \cdot 2H_2O + 2SO_3 \rightarrow HClO_4 + 2H_2SO_4$$

More recently, perchloric acid has been made by distillation from a mixture of aqueous (72–85%) perchloric acid and magnesium perchlorate, a powerful dessicant.[3] Distillation under reduced pressures between 20° and 70°C gives an 80% yield of the anhydrous acid. 73·6% perchloric acid remains in the still at the end of the distillation.

There are certain necessary precautions associated with the preparation of anhydrous perchloric acid. It is a powerful and very reactive oxidant and silicone greases and other lubricants explode on contact with it and cannot be used for ground-glass joints in the preparative apparatus. The 72·5% aqueous acid has been recommended as an alternative lubricating agent although joints sealed in this way have been found[4] not to hold a vacuum. The anhydrous acid reacts explosively with materials like wood, paper, carbon and organic solvents and so must never be allowed to come into contact with these.

Anhydrous perchloric acid boils undecomposed at 16°C at a pressure of 2·4 kNm^{-2} but cannot be distilled without decomposition at atmospheric pressure. Its extrapolated b.p. at atmospheric pressure is 130° and its m.p. = −102°C. If it is not used within 12 to 24 hr of preparation, there is, at room temperature, a gradual accumulation of decomposition products. When this occurs, it may explode spontaneously. Pure samples do not explode when stored for about one month at ordinary temperatures and they may be stored for up to two months at liquid–air temperatures without significant decomposition. When the anhydrous acid is required it is best prepared fresh from the monohydrate by distillation of this at reduced pressure and at 110°C:

$$2HClO_4 \cdot H_2O \rightarrow HClO_4 + HClO_4 \cdot 2H_2O$$

The most important commercial process for the preparation of perchlorate and perchloric acid is the anodic oxidation of chlorate to perchlorate. In the Kreider-Mathers process for the manufacture of perchloric acid, sodium chloride is the raw material. A saturated solution of this is electrolysed with a graphite anode and an iron cathode and about 85% of the sodium chloride is converted to chlorate. Unchanged chloride is removed by concentrating the electrolyte so that it crystallizes out. The next stage is the electrolysis of a saturated solution of sodium chlorate with a platinum anode and an iron cathode. More than 99% of the chlorate can be oxidized to perchlorate. The sodium perchlorate is converted to perchloric acid by metathesis with concentrated hydrochloric acid, the precipitated sodium chloride removed by centrifugation and the aqueous perchloric acid distilled under vacuum to provide the purified product. The sodium chloride and hydrochloric acid are recovered for further processing.

125

Reactions

Perchloric acid is an extremely strong acid and contains more oxygen than any other common acid. Its oxidizing power depends very markedly on both its concentration and the temperature. The anhydrous acid reacts vigorously and in many cases explosively at room temperature with reducing agents whereas aqueous solutions of perchloric acid up to 73·6% do not show oxidizing properties in the cold. Hot concentrated perchloric acid (70–73·6%) is a powerful oxidizing agent, decomposing to give chlorine, oxides of chlorine and oxygen as reduction products. The perchlorate ion shows no oxidizing properties in aqueous solution. Most metallic perchlorates are very soluble in water: exceptions are moderately soluble potassium, rubidium and caesium salts. Perchloric acid forms sparingly soluble salts with many organic bases and this is one of the most important classical applications of the acid for the isolation and identification of such compounds.

The reactions of anhydrous perchloric acid with metal chlorides and nitrates have been studied.[4] Reaction with iron(III) chloride gives the hydrated perchlorate, not the anhydrous salt. Formation of the hydrate is attributed to existence of the equilibrium

$$2HClO_4 \rightleftharpoons Cl_2O_7 + H_2O$$

in anhydrous perchloric acid. The reaction with anhydrous metal nitrates proceeds according to the general equation:

$$MNO_3 + 3HClO_4 \rightarrow MClO_4 + NO_2 \cdot ClO_4 + H_3O \cdot ClO_4$$

This reaction goes to completion with potassium nitrate but is only 80–90% complete with transition metal nitrates. These abstract the elements of water from perchloric to give partially hydrated perchlorates.

Molecular Structure of Anhydrous Perchloric Acid

Two electron diffraction studies on the vapour have been made.[5,6] The bond lengths and angles found by Clark and co-workers are shown in Fig. 13.1. The ClO_4 group has

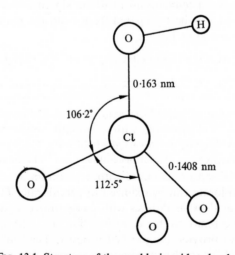

Fig. 13.1. Structure of the perchloric acid molecule.

C_{3v} symmetry. The exact position of the hydrogen is not known. The Cl–O(H) bond length is significantly shorter than the value 0·170 nm found for Cl–O in Cl_2O_7 and Cl_2O and generally regarded as that of a single bond. There appears to be some π-bond character in Cl–O(H). The other bond lengths are appreciably shorter still and signify that these bonds have pronounced multiple character.

Hydrates of Perchloric Acid

Several hydrates of perchloric acid are known to exist. The classical study is that of van Wyk[7] who concluded from melting-point data that five hydrates, containing respectively 1, 2, 2·5, 3 and 3·5 moles of water per mole of perchloric acid, could exist. He suggested that the anhydrous acid is always associated with chlorine heptoxide and the monohydrate according to:

$$3HClO_4 \rightleftharpoons Cl_2O_7 + HClO_4 \cdot H_2O$$

The monohydrate is the most well known of the hydrates. It is a colourless crystalline compound, m.p. 49·90°C. Its physical and chemical properties have been thoroughly studied. Mascherpa and co-workers[8] have confirmed the existence of the four other hydrates within the range of 55 to 75% aqueous perchloric acid. Their phase diagram was determined by a combination of differential thermal analysis and solubility measurements and is illustrated in Fig. 13.2. E_A, E_B, E_C and E_D are eutectic mixtures. Conflicting views have been expressed over the existence of the equilibrium expressed in the above equation. Zinov'ev and Rosolovskii[9] have concluded, from thermal analysis and melting-point studies on the $Cl_2O_7(H_2O)$ system between 25 and 100 mole%, that a eutectic, m.p. $= -100 \pm 2$°C, exists at 53 mole% Cl_2O_7. This could be composed of the monohydrate and either Cl_2O_7 or a metastable compound, $Cl_2O_7 \cdot 2HClO_4$. Trowbridge and Westrum[10] have suggested that this eutectic does not exist but that pure perchloric acid

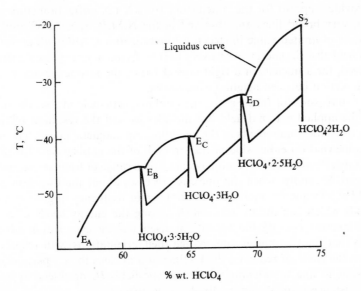

FIG. 13.2. Phase diagram showing the interrelationships between perchloric acid hydrates.

does as a single species in the crystalline state. For example, the vibrational spectrum[11] of anhydrous perchloric acid shows neither the intense band at 931 cm^{-1} due to $HClO_4 \cdot H_2O$ nor the bands at 501 and 695 cm^{-1} expected for Cl_2O_7. Neither of these species can therefore be present in significant amounts.

Structure of Perchloric Acid Monohydrate

The monohydrate is isomorphous with ammonium perchlorate[12] and should therefore be formulated as oxonium perchlorate, $H_3O^+ClO_4^-$. X-ray diffraction[13] has shown the crystal is orthorhombic and that the unit cell contains four formula units and has the dimensions: $a = 0.9065$, $b = 0.5569$ and $c = 0.7339$ nm. The average Cl,O bond length is 0.142 ± 0.001 nm. The presence of tetrahedral ClO_4^- ions is shown by the equality of all Cl,O bond lengths. If $HClO_4$ molecules were present, one bond would be noticeably longer than the other three. This also establishes the presence of H_3O^+ ions in the crystal. There appears to be some disorder in the orientation of these from the X-ray data and this is confirmed by N.M.R. studies discussed later.

The above value of the Cl,O bond length is significantly shorter than the value of 0.146 ± 0.001 nm found[14] in nitronium perchlorate, $NO_2^+ClO_4^-$. This difference has been attributed[15] to the omission of a correction for the rotational oscillation of the perchlorate ion in the acid hydrate. When such a correction is applied, the mean value of the Cl,O bond length becomes 0.1452 ± 0.0005 nm.

Another crystalline modification of perchloric acid monohydrate exists at low temperatures,[16] the transition temperature between the two forms being around $-30°C$. The form stable at low temperatures is monoclinic, the unit cell contains four formula units and has the dimensions: $a = 0.7541$, $b = 0.9373$, $c = 0.5359$ nm and $\beta = 97°41'$. The average Cl,O bond length $= 0.1464$ nm. The H_3O^+ ion is pyramidal, the angle H–O–H $= 112°$ and the H...H distance $= 0.14$ nm.

The proton magnetic resonance spectra of the solid monohydrate at different temperatures provides support for the conclusions from X-ray diffraction studies.

In general, very broad lines are observed in the N.M.R. spectra of solid samples and such spectra are of limited value in structural investigations. Although an isolated nucleus in a uniform field absorbs energy over a very narrow frequency range and a sharp resonance line is observed, for a nucleus in a rigid crystal lattice the resonance line is broadened by spin–spin interaction between neighbouring nuclei.

There are two reasons for this. Firstly, the total magnetic field at a single nucleus consists not only of the applied external field but includes as well the resultant of the local fields produced by the static components of the neighbouring magnetic nuclei. The total magnetic field varies somewhat on either side of the strength of the applied field for different nuclei at any given time. Resonance absorption occurs over a range of frequencies and a broadened line is obtained. Secondly, when the nucleus precesses about the direction of the applied magnetic field at a frequency equivalent to the Larmor frequency, it gives rise to oscillatory magnetic fields which can induce transitions among the energy levels of a neighbouring nucleus. The nucleus causing the transitions itself undergoes a simultaneous change in spin state, so that the two nuclei interchange their orientations with respect to the main field. This has the effect of reducing the lifetime of a nucleus in any particular spin energy level and results in line broadening. The magnetic field, H, produced at a neighbouring nucleus at a distance τ by a nuclear dipole, μ, is given by

$$H = \pm\mu\tau^{-3}(3\cos^2\theta - 1)$$

and so the width of an absorption line is related to the third power of the internuclear distance. H is also a function of θ, the angle between the line joining the two nuclei to the direction of the applied magnetic field. The resonance frequencies observed in the spectrum of a single crystal will change in position as the orientation of the crystal changes with respect to the applied magnetic field.

At $-183°C$, the N.M.R. curve for the monohydrate shows three peaks but at $0°C$, only a single narrow absorption line is observed[17] (Fig. 13.3). The line shape at the lower of these

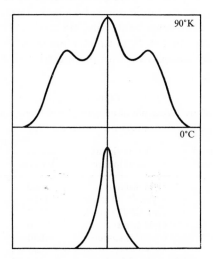

Fig. 13.3. Nuclear magnetic resonance spectrum of perchloric acid monohydrate at $-183°C$ and $0°C$.

temperatures is characteristic of a triangular configuration of identical nuclei. It has been predicted[18] from theoretical considerations that the N.M.R. spectrum of a solid containing an ion or molecule of this shape is composed of a central absorption line, with probability 0·5, flanked by two lines, one on either side, with probability 0·25. The spectrum measured at $-183°C$ is therefore interpreted in terms of protons located at the corners of an equilateral triangle in the oxonium ion, H_3O^+. The H,H distance within this ion is $0·172 \pm 0·002$ nm. It is not possible to decide, on this evidence alone, whether H_3O^+ is flat or pyramidal. The spectrum at $0°C$ is caused by rapid reorientations or rotations of the oxonium ions probably occurring about more than one axis. These data and conclusions have been substantiated in an independent investigation of the N.M.R. spectrum of the monohydrate.

Structure of Perchloric Acid Dihydrate

Perchloric acid dihydrate melts at $-17·8°C$. The solution of the dihydrate corresponds to 73·60 wt. % perchloric acid and is slightly more concentrated than the azeotropic mixture. A solution of this composition can be obtained by vacuum distillation of the azeotrope. The crystal structure at $-190°C$ has been determined[21] from single-crystal X-ray data. The crystals are orthorhombic, with four formula units in a cell of dimensions: $a = 0·5819$, $b = 1·0598$, $c = 0·7369$ nm. The water molecules are bound to each other in pairs by a very short hydrogen bond (0·2424 nm in length) forming diaquohydrogen ions,

$H_5O_2^+$. These ions are hydrogen-bonded to perchlorate ions into layers which are held together by van der Waals forces. The average Cl,O distance in the perchlorate ion is 0·1438 nm. $H_5O_2^+$ ions have been identified in some other acid hydrates, such as HCl · $2H_2O$ and HCl · $3H_2O$.

Vibrational Spectra

The infrared spectra of anhydrous $HClO_4$ and $DClO_4$ have been measured[22] between 300 and 5000 cm^{-1} in the gaseous, liquid and solid states. Twelve fundamental frequencies are expected for a covalent molecule containing six atoms. These are symmetric (A') or asymmetric (A'') with respect to the C_s symmetry plane of the molecule. All have been assigned (Table I), partly by comparison with the spectrum of $FClO_3$, a molecule spectro-scopically similar to $HClO_4$.

TABLE I

Infrared spectrum of anhydrous perchloric acid

Class	Number of vibration	Approximate description	Frequencies (cm^{-1}) Vapour	Solid	Liquid
A'	ν_1	O–H stretch	3560	3260	3275
A''	ν_9	asym. Cl–O stretch	1326	1315	1315
A'	ν_2	sym. Cl–O stretch	1263	1283	
A'	ν_3	O–H bend	1200	1245	1245
A'	ν_4	Cl–O(H) stretch	1050	1033	1033
A'	ν_4	^{37}Cl–O(H)	1047		
A'	ν_5	Cl–O(H) stretch	725	740 760	743
A'	ν_5	^{37}Cl–O(H)	721		
A''	ν_{10}	O–Cl–O bend	579	585	582
A'	ν_6	O–Cl–O bend	560	566	571
A'	ν_7	O–Cl–O bend	519		
A''	ν_{11}	O–Cl–O(H) bend	450 430	428	440
A'	ν_8	O–Cl–O(H) bend	413 390	371	
A'	ν_{12}	torsion OH	318 307	478	480

In the vapour phase, the O–H stretching frequency appears at 3560 cm^{-1}. In the liquid and solid, the band due to this vibration is found at a significantly lower frequency because of hydrogen-bonding in the condensed phase. The difference is less than that observed between the gas and condensed phases for other simple molecules such as H_2SO_4, HNO_3, CH_3OH and H_2O and indicates that the hydrogen bonds in liquid and solid perchloric acid are relatively weak. Their strength has been estimated at 12·5 kJ mol^{-1} and the weak molecular association is in accordance with the low viscosity and high volatility of the liquid and the low melting-point of the solid.

The two modifications of the solid monohydrate have been characterized[23] by their Raman spectra. The orthorhombic form, stable above $-30°C$, shows bands due to per-chlorate and oxonium ions but none due to any other species. The orthorhombic lattice has low site symmetry and there is some hydrogen-bonding between the ions. Nevertheless, there is no splitting of the fundamental vibrations of the perchlorate ion which should follow from its location in a site of reduced symmetry. It appears there is some structural disorder, for example, hindered ionic rotation of the type indicated by N.M.R. spectroscopy which accounts for the absence of any splitting. The Raman spectrum measured at $-50°C$ is different from that measured at $-20°C$ and above and this is consistent with a phase change occurring in the monohydrate at about $-30°C$.

Ionic Rotation in $HClO_4 \cdot H_2O$

The energy barrier to rotation of the oxonium ion has been estimated from measurements[24] of the total neutron scattering cross-section for $HClO_4 \cdot H_2O$. This method has been used for the determination of the barrier to rotation of NH_4 groups in crystalline ammonium compounds on the basis of the correlation which exists between the scattering of low-energy neutrons and the rotational freedom of the ammonium ions. At neutron energies far below thermal, the neutron scattering cross-section is proportional to the neutron wavelength, and the slope of the plot of cross-section against wavelength increases with the freedom of motion of the hydrogen atoms. For example, the slope is significantly greater for NH_4ClO_4 than for $(NH_4)_2Cr_2O_7$ and is consistent with almost free rotation[25] of the ammonium ion in NH_4ClO_4. The value of the slope for a freely-rotating pyramidal molecule such as CH_3I is 15·7 am. The slope is only 8 am for crystalline $HClO_4 \cdot H_2O$ and this confirms the conclusion from its vibrational spectrum that there is hindered rotation of the H_3O^+ ions in the solid. The energy barrier to rotation is estimated as about 7·5 kJ mol^{-1}. For comparison, the slope for ammonium perchlorate is 13 am and the corresponding energy barrier only 0·42 to 0·84 kJ mol^{-1}.

Dissociation of Perchloric Acid

The dissociation of perchloric acid in aqueous solutions has been studied by a variety of techniques, particularly nuclear magnetic resonance and Raman spectroscopy. The basis of the N.M.R. studies is that the proton resonance shift in an aqueous solution of any acid is a colligative property. Experimentally, it is found that the proton resonance of perchloric acid in water is a single peak. One might expect to observe a number of individual proton resonance lines at positions corresponding with different proton-containing ions and molecules. This is not so because fast exchange of protons occurs between these various species. Instead a single resonance is found which corresponds with the concentration-weighted average of the resonance of the proton in different environments.

In N.M.R. studies,[26] the resonance shift, s, is found from the applied fields, H_0 and H, necessary to cause resonance in pure water and the solution respectively. s is defined by:

$$s = \left(\frac{H_0}{H} - 1\right) 10^6 + g$$

$g = \left(\frac{2\pi}{3}\right)(\chi - \chi_0) 10^6$ is a correction factor allowing for the difference in volume magnetic susceptibility of the solution, χ, compared with that of pure water, χ_0. s varies with p, the stoichiometric mole fraction of hydrogen ion in H_3O^+ on a total hydrogen basis and defined by $p = \dfrac{3x}{2 - x}$ where x is the stoichiometric mole fraction of the acid. When $HClO_4 : H_2O = 1:1$, $x = \frac{1}{2}$ and $p = 1$. For the anhydrous acid, $x = 1$ and $p = 3$.

The degree of dissociation, α, is given by the equation

$$\frac{s}{p} = s_1\alpha + \frac{s_2}{3}(1 - \alpha)$$

in which s_1 = shift due to the oxonium ion and s_2 = shift due to the undissociated acid, assumed to be equal to that for the anhydrous acid. s_1 is the limiting slope of the plot of s against p as p tends to zero. Hood and co-workers[26] have estimated that the equilibrium constant of the reaction:

$$\underset{(1-\alpha)}{HClO_4} + \underset{(\alpha)}{H_2O} \rightleftharpoons \underset{(\alpha)}{H_3O^+} + \underset{(\alpha)}{ClO_4^-}$$

is 38. The degree of dissociation is about 0·8 in 11 mol dm^{-3} and 0·9 in 8 mol dm^{-3} perchloric acid at room temperature.

Several Raman studies of the dissociation of perchloric acid have been made. The perchlorate ion has tetrahedral symmetry, T_d, and its Raman spectrum in solution is characterized[27] by four lines corresponding to two stretching and two bending vibrations. The intensity of the line observed at 931 cm^{-1} serves as a measure of the concentration of perchlorate ion and hence of the extent of dissociation of the acid. The spectrum of the anhydrous acid shows eight lines. Seven of these arise from the reduced symmetry of the ClO_4 group in the $HClO_4$ molecule of trigonal symmetry, C_{3v}. The eighth is an O–H stretching vibration.

From the first Raman study[28], it was concluded that perchloric acid is completely dissociated in aqueous solutions of concentration 9 mol dm^{-3} and less. This estimate is significantly different from that derived from N.M.R. measurements and the discrepancy has prompted further investigations.

There is no indication of perchlorate ions in the pure acid but, on dilution with water, the most intense perchlorate Raman line (that at 931 cm^{-1}) is already apparent in 97% w/w acid.[28] Further dilution results in further dissociation and the acid is completely dissociated at 77% composition.

In another study,[29] α has been found to be unity for concentrations up to 8 mol dm^{-3}, decreasing above this to 0·88 at 12 mol dm^{-3}. It appears to be definitely established[30] that α does not differ significantly from unity until the molarity of the aqueous acid is greater than about 10 mol dm^{-3}. This behaviour is related to the extent to which a proton can be solvated. In dilute solution, it is believed to exist as $H_9O_4^+$ ions. Above 10 mol dm^{-3}, there is insufficient water present to solvate, in this way, all the protons which would result from the complete dissociation of perchloric acid. Hence α drops below one.

The differences between values reported for the degree of dissociation of perchloric acid in aqueous solution may arise from a breakdown of one or more of the assumptions on which the interpretation of the Raman or N.M.R. data is based or from some inherent limitation in the instrumental method. The assumptions underlying the calculations from proton chemical shifts have particularly attracted criticism. The chemical shifts of the proton in H_3O^+ and $HClO_4$, s_1 and s_2 respectively, are regarded as invariant and independent of concentration. The assumption that s_1 is constant is difficult to justify in view of the decreasing extent of hydration of H^+ which must accompany increasing acid concentration. s_2, the chemical shift for undissociated perchloric acid molecules in aqueous solution, is taken as the value measured in anhydrous perchloric acid. It is not necessarily correct to do this in view of the greater extent to which intermolecular hydrogen-bonding should occur in the pure acid.

The ^{35}Cl N.M.R. is less susceptible than proton resonance to environmental influences. Recent measurements[31] of ^{35}Cl resonance shifts in aqueous perchloric acid have confirmed the Raman results that α is unity at least up to 8 mol dm^{-3} and is just below 0·9 at 12 mol dm^{-3}.

The main differences between the results obtained by various workers is that ^{35}Cl resonance and Raman measurements indicate a much higher value for the extent of dissociation at high concentrations than do the proton resonance measurements. This is probably because the chemical shift of the hydrated proton changes with concentration. There is general agreement that perchloric acid is completely dissociated up to at least 6 to 7 mol dm^{-3}.

Acidity Function of Perchloric Acid

The extensive dissociation of perchloric acid in aqueous solution even at high concentrations places it in a special position among acids. According to the Bronsted definition, an acid is a substance which tends to donate a proton to a base. For the acid, HA, the equilibrium established in aqueous solution is:

$$HA + H_2O \rightleftharpoons H_3O^+ + A^-$$

where the solvent behaves as the base. The magnitude of the equilibrium constant is a measure of the strength of the acid. Perchloric acid is classed as a strong acid because of the large value of its dissociation constant in concentrated solution and complete dissociation in dilute solution. The common mineral acids, HCl, HNO$_3$ and H$_2$SO$_4$ are also completely ionized in dilute aqueous solution and are similarly defined as strong. In some respects, the nature of the anion in a dilute solution of a strong acid is therefore of relatively small significance.

In more concentrated solutions, differences in the behaviour of strong acids become more marked and it is valuable to have a quantitive means of comparing acid strengths. To do this, Hammett and Deyrus[32] proposed the acidity function, H_0, defined by

$$H_0 = -\log \{H^+\} f_B / f_{BH^+}$$

where $\{H^+\}$ = activity of H^+ and f_B and f_{BH^+} are the respective activity coefficients of a base B and its protonated form BH$^+$. In dilute solution, the activity coefficient ratio becomes unity and $H_0 = $ pH. Also

$$H_0 = \log \frac{[B]}{[BH^+]} + pk_{BH^+}$$

where

$$pk_{BH^+} = -\log \frac{\{H^+\} \{B\}}{\{BH^+\}}$$

If B is an indicator, the ratio $\frac{[B]}{[BH^+]}$ can be determined spectrophotometrically and pk_{BH^+} calculated by conventional methods.

For two bases, B$_1$ and B$_2$, their pk values can be compared by the relationship:

$$pk_1 - pk_2 = -\log \frac{[B_1][B_2H^+]}{[B_1H^+][B_2]} - \log \frac{f_{B_1} f_{(B_2H^+)}}{f_{(B_1H^+)} f_{B_2}}$$

On the assumption that the ratio f_B/f_{BH^+} is the same for all bases and depends only on the acidity, this equation reduces to

$$pk_1 - pk_2 = -\log \frac{[B_1]\,[B_2H^+]}{[B_1H^+]\,[B_2]}.$$

H_0 has a value independent of the particular indicator used to the extent that this fundamental assumption is correct. A series of organic bases, derived from nitroaniline, was used by Hammett and many other investigators since to determine the acidity functions of various strong acids. In the case of perchloric acid, values up to 78·6% by weight have been determined.[33] These are compared in Fig. 13.4 with data for sulphuric acid.[34] Increasingly

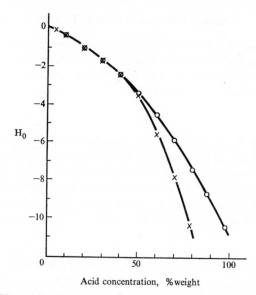

FIG. 13.4. The acidity functions of perchloric acid (x—x) and sulphuric acid (o—o).

negative values of H_0 correspond to higher acidities and 78·6% perchloric acid has an acidity, measured by reference to organic indicators, equivalent to that of 98% sulphuric acid. For a particular mole fraction of acid, the acidity function of perchloric acid is always more negative than that of other inorganic acids such as HCl, HF, HNO_3 and H_3PO_4. Perchloric acid is therefore regarded as one of the strongest acids known. This conclusion receives confirmation from studies using a differentiating solvent such as pyridine. The relative dissociation constants of some common acids when determined[35] by potentiometric titrations in pyridine follow the order:

$$HClO_4 > HI > HNO_3 > HBr > HCl$$

The acid is widely used as a titrant in non-aqueous media. When used in a solvent like acetic acid, it becomes possible to carry out the potentiometric titration of a variety of weak bases not easily titrated in other solvents. Bases which are weak in aqueous solution become stronger in acetic acid and more pronounced breaks in the potentiometric titration curves are observed using this solvent. In acetic acid solutions of perchloric acid, the acidic species is $H_2OOC \cdot CH_3^+ClO_4^-$. When acetic anhydride is added to acetic acid as

134

solvent, even weaker bases, such as amides, may be successfully titrated[36] using perchloric acid. The reason for this is the formation of another acidic species:

$$CH_3COOH_2^+ + (CH_3CO)_2O \rightleftharpoons (CH_3CO)_2OH^+ + CH_3COOH$$

$$(CH_3CO)_2OH^+ \rightleftharpoons CH_3CO^+ + CH_3COOH$$

The titration of an amide then proceeds according to:

$$R \cdot \overset{\displaystyle O}{\overset{\displaystyle \|}{C}}-NH_2 + CH_3CO^+ClO_4^- \rightleftharpoons R \cdot \overset{\displaystyle O}{\overset{\displaystyle \|}{C}} \cdot NH_2 \cdot CH_3CO^+ClO_4^-$$

References

1. Von Stadion, F., *Gilb. Ann. Physik*, **22**, 197 (1816).
2. Roscoe, H. E., *J. Chem. Soc.* **16**, 82 (1863).
3. Smith, G. F., *Talanta* **7**, 212 (1961).
4. Hathaway, B. J., and Underhill, A. E., *J. Chem. Soc.* 648 (1960).
5. Akishin, P. A., Vilkov, L. V., and Rosolovskii, V., *Krist.* **4**, 353 (1959).
6. Clark, A. H., Beagley, B., and Cruickshank, D. W. J., *Chem. Comm.* 14 (1968).
7. Van Wyk, H. J., *Z. anorg. Chem.* **48**, 1 (1906).
8. Mascherpa, G., Pavia, A., and Potier, A., *Compt. Rend.* **254**, 3213 (1962).
9. Zinov'ev, A. A., and Rosolovskii, V. Y., *Zh. Neorgan. Khim.* **3**, 2382 (1958).
10. Trowbridge, J. C., and Westrum, E. F., *J. Phys. Chem.* **68**, 42 (1964).
11. Dahl, A. J., Trowbridge, J. C., and Taylor, R. C., *Inorg. Chem.* **2**, 654 (1963).
12. Volmer, M., *Ann.* **440**, 200 (1924).
13. Lee, F. S., and Carpenter, G. B., *J. Phys. Chem.* **63**, 279 (1959).
14. Truter, M. R., Cruickshank, D. W. J., and Jeffrey, G. A., *Acta Cryst.* **13**, 855 (1960).
15. Truter, M. R., *Acta Cryst.* **14**, 318 (1961).
16. Nordman, C. E., *Acta Cryst.* **15**, 18 (1962).
17. Richards, R. E., and Smith, J. A. S., *Trans. Faraday Soc.* **47**, 1261 (1951).
18. Andrew, E. R., and Bersohn, R., *J. Chem. Phys.* **18**, 159 (1950).
19. Kakiuchi, Y., Shono, H., Komatsu, H., and Kigoshi, K., *J. Phys. Soc. Jap.* **7**, 102 (1952).
20. Kakiuchi, Y., and Komatsu, H., *J. Phys. Soc. Jap.* **7**, 380 (1952).
21. Olovsson, I., *J. Chem. Phys.* **49**, 1063 (1968).
22. Giguere, P. A., and Savoie, R., *Can. J. Chem.* **40**, 495 (1962).
23. Taylor, R. C., and Vidale, G. L., *J. Amer. Chem. Soc.* **78**, 5999 (1956).
24. Janik, J., *Acta Phys. Polon.* **27**, 491 (1965).
25. Rush, J. J., Taylor, T. I., and Havens, W. W., Jr., *J. Chem. Phys.* **35**, 2265 (1961).
26. Hood, G. C., Redlich, O., and Reilley, C. A., *J. Chem. Phys.* **22**, 2067 (1954).
27. Redlich, O., Holt, E. K., and Bigeleisen, J., *J. Amer. Chem. Soc.* **66**, 13 (1944).
28. Simon, A., and Weist, M., *Zeit. Anorg. Allgem. Chem.* **268**, 301 (1952).
29. Heinzinger, K., and Weston, R. E., Jr., *J. Chem. Phys.* **42**, 272 (1965).
30. Covington, A. K., Tait, M. J., and Wynne-Jones, W. F. K., *Proc. Roy. Soc. A* **286**, 235 (1965).
31. Akitt, J. W., Covington, A. K., Freeman, J. G., and Lilley, T. H., *Chem. Comm.* 349 (1965).
32. Hammett, L. P., and Deyrus, A. J., *J. Amer. Chem. Soc.* **54**, 2721 (1932).
33. Bonner, T. G., and Lockhart, J. C., *J. Chem. Soc.* 2840 (1957).
34. Yates, K., and Wai, H., *J. Amer. Chem. Soc.* **86**, 5408 (1964).
35. Mukherjee, L. M., and Kelly, J. J., *J. Phys. Chem.* **71**, 2348 (1967).
36. Pietrzyk, D. J., *Anal. Chem.* **39**, 1367 (1967).

CHAPTER 14

XENON TETRAFLUORIDE

Introduction

The noble gases were discovered towards the end of the nineteenth century when the researches of Lord Rayleigh and Sir William Ramsey showed that the five elements, helium, neon, argon, krypton and xenon, all occur in the earth's atmosphere. These elements were characterized by complete chemical inertness. Early claims that some compounds of these gases existed could never be substantiated when subsequent attempts were made to repeat the syntheses. The only exception is the group of 'clathrate' compounds produced by interaction between a noble gas and water or various organic compounds such as methanol, chloroform, acetic acid and quinol. In these, the noble gas atoms are trapped within the cage-like structure of the host lattice. There is no chemical bonding between components which involves the electrons of the noble gas atom, the only forces being of the van der Waals type.

In 1933, Yost and Kaye[1] reported unsuccessful attempts to prepare a fluoride of xenon by passage of an electric discharge through a mixture of fluorine and xenon gases. Their work gave no definite evidence for compound formation but they did not exclude the possibility that xenon fluorides could exist. Difficulties in handling fluorine and a limited supply of xenon must have been contributory factors in their failure to produce noble gas compounds under experimental conditions so close to those now successfully used for such preparations.

For many years the inert character of the noble gases was believed to demonstrate a particularly high stability associated with a complete octet of electrons in the valence shell of an atom. This formed the basis of a widely accepted rule of valence theory, namely that the electronic structure of many molecular and ionic compounds can be simply described in terms of the constituent atoms sharing, gaining or losing a number of electrons sufficient to acquire the electronic arrangements of a noble gas. It has been recognized that this rule is of only limited application for there are many stable molecules and ions in which the atoms have more than eight electrons in their valence shells. There is no reason why a noble gas atom should not form conventional chemical bonds in the same way as any other element.

An authentic compound of the noble gas, xenon, was first prepared by Bartlett in 1962.[2] One of the products of the reaction of platinum salts with fluorine in glass and silica apparatus is a deep-red solid, which can also be made by the reaction between molecular oxygen and platinum hexafluoride at room temperature. This is formulated as $O_2^+PtF_6^-$ and, because the ionization potentials of molecular oxygen and xenon are almost identical, the synthesis of the xenon analogue, $Xe^+PtF_6^-$, was attempted and achieved.

The synthesis of other xenon compounds quickly followed and a number are now known in which xenon is combined with fluorine, chlorine or oxygen. The existence of compounds of krypton and radon has also been established. Many physical methods have been used to examine noble gas compounds in order to determine their structure and the nature of the bonding therein. As might be expected, the problem of the description of bonds in these novel compounds has proved a stimulating challenge to theoretical chemists.

Preparation

Xenon tetrafluoride was the first simple xenon compound to be synthesized.[3] Xenon and fluorine were mixed together in a volume ratio of 1:5 in a sealed nickel can and heated to 400°C for 1 hr. The can was cooled quickly to −78°C and excess fluorine gas pumped off. The main product of this reaction was the tetrafluoride, XeF_4, although smaller amounts of XeF_2 and XeF_6 were also formed. XeF_4 is a colourless solid which is stable at room temperature when dry and free of impurity.

Direct reaction between xenon and fluorine proceeds in a step-wise fashion to give a mixture of fluorides. No product is obtained unless the temperature is above 120°C indicating the necessity to reach a certain minimum concentration of fluorine atoms for any reaction with xenon to occur.

The formula of xenon tetrafluoride was simply established in two ways. The increase in weight of the contents of the nickel can when compared with the weight of xenon taken showed that the atomic ratio F:Xe was close to 4:1. Also xenon tetrafluoride is quantitatively converted to hydrogen fluoride by heating in fluorine. The composition of the xenon compound was confirmed by volumetric titration of the HF formed.

Other methods for the preparation of XeF_4 also give an impure product. It is formed[4] when small amounts of xenon are introduced into a fluorine stream passing slowly through a reaction tube heated to between 300° and 500°C. It has been successfully synthesized[5] by passage of an electric discharge through a mixture of xenon and fluorine (volume ratio = 1:2) in a reaction vessel cooled to −78°C. In a typical run, virtually quantitative conversion to XeF_4 was achieved in 3·5 hr. The method has been recommended as a means of production of XeF_4 on a continuous basis.

Full details for a laboratory-scale preparative method have been given.[6] Special precautions are necessary to prevent the formation of xenon trioxide. This could result from the hydrolysis of XeF_4 and then dehydration of the hydrolysis product. XeO_3 is dangerously explosive and so a rigorous exclusion of water from apparatus used in the preparation of XeF_4 is necessary.

An ingenious purification procedure for XeF_4 has been devised[7] to separate it from XeF_2 and XeF_6. It is based on the ability of XeF_2 and XeF_6 to donate fluoride ion and form ionic adducts with AsF_5 whereas XeF_4 does not behave in this way. The mixture of XeF_2, XeF_4 and XeF_6 is dissolved in BrF_5 and the solution treated with excess AsF_5. The solvent and excess BrF_5 are removed under vacuum at 0°C leaving behind the adducts $[Xe_2F_3^+][AsF_6^-]$ and $[XeF_5^+][AsF_6^-]$ and unchanged XeF_4. Pure XeF_4 is recovered by sublimation at 20°C under vacuum and collected in a trap cooled to −60°C.

137

Reactions

Xenon tetrafluoride is a moderately strong fluorinating agent and it has been compared, in this respect, to UF_6. It reacts with excess xenon at 400°C to give XeF_2. It slowly converts SF_4 to SF_6 at room temperature. Like XeF_2, it fluorinates ethylene to give 1,1- and 1,2-difluoroethane and propylene to give 1,1-difluoropropane as the chief product.

Hydrolysis proceeds according to:

$$3XeF_4 + 6H_2O \rightarrow XeO_3 + 2Xe + 1.5O_2 + 12HF$$

A third of the xenon is retained in solution as XeO_3, a strongly oxidizing compound which, for example, converts HCl to Cl_2. Optimum yields of XeO_3 are obtained by hydrolysis in acid or neutral solution: in basic media, Xe(VI) disproportionates to Xe gas and Xe(VIII) (perxenate).

Analytically, XeF_4 can be determined by its reactions with hydrogen,

$$XeF_4 + 2H_2 \rightarrow Xe + 4HF$$

with mercury,

$$XeF_4 + 4Hg \rightarrow Xe + 2Hg_2F_2$$

and with iodides,

$$XeF_4 + 4KI \rightarrow Xe + 2I_2 + 4KF$$

The xenon released can be measured in each case, HF and iodine can be determined volumetrically and Hg_2F_2 can be determined gravimetrically.

Structure

Several diffraction studies of xenon tetrafluoride were reported soon after it was discovered. These were three structural analyses by X-ray diffraction,[8-10] one by neutron diffraction[11] and one by electron diffraction.[12] As X-rays are scattered by the electrons of atoms in the crystal and the scattering amplitude of an atom is determined by the number of its electrons, the xenon atom scatters electrons to a much greater extent than the fluorine atoms. This makes it difficult to locate the exact positions of the fluorines by X-ray diffraction and it is inherently more precise to study a solid like XeF_4 by neutron diffraction. Neutrons are scattered chiefly by atomic nuclei and there is little difference between the scattering cross-section for neutrons of xenon and fluorine. The main disadvantage of neutron diffraction is that relatively much longer exposure times are needed to collect intensity data than are required in X-ray diffraction work. This disadvantage can be largely offset by the use of automatic, computer-controlled apparatus for neutron diffraction studies.

Results of the various X-ray diffraction studies are in substantial agreement with one another. The unit cell is monoclinic and contains two planar XeF_4 molecules. The data of Templeton, Zalkin, Forrester and Williamson[9] are recorded in Table I. These investigators reported two internuclear Xe,F distances as 0.190 and 0.192 nm. As the standard deviation estimated for their measurements was 0.002 nm, it appears there is no significant difference between these, and all Xe,F bonds are the same within the limits of experimental error. All the atoms in the crystal are in thermal motion, with the light fluorine atoms moving more than the heavy xenon atoms. Because of this, the average bond length, that is, the mean separation of Xe and F atoms, is believed to be slightly greater than the values quoted and is taken to be 0.193 nm.

138

TABLE I

Molecular Parameters for XeF_4 determined by X-ray and Neutron Diffraction

	X-ray diffraction	Neutron diffraction
Unit cell	Monoclinic	Monoclinic
Space group	$C_{2h}^5 - P2_{1/n}$	$P2_{1/n}$
Unit cell dimensions	$a = 0.505$;	
	$b = 0.592$;	
	$c = 0.577$ (in nm)	
	$= 99.6 \pm 0.1°$	
Xe–F_1 (nm)	0.192 ± 0.002	$0.1954 \pm 0.0002*$
Xe–F_2 (nm)	0.190 ± 0.002	$0.1951 \pm 0.0002*$
F–Xe–F angle (deg)	90.4 ± 0.9	90.0 ± 0.1

* Bond lengths corrected for thermal motion of the atoms.

Molecular parameters determined by neutron diffraction[11] are also given in Table I. These results agree with those from X-ray work and both confirm the conclusion, from the electron diffraction study, that XeF_4 is a square planar molecule with D_{4h} symmetry.

Vibrational Spectra

For a square planar molecule belonging to the point group D_{4h}, there are nine fundamental vibrational modes. These are distributed among the symmetry species of this group as follows: $A_{1g} + B_{1g} + B_{2g} + A_{2u} + B_{2u} + 2E_u$ and they are illustrated in Fig. 14.1. This notation follows from the choice of the directions of the C_2' and C_2'' symmetry axes to be coincident with the axes of two mutually perpendicular Xe–F bonds. Then the asymmetric stretching vibration, ν_5, is designated as B_{1g}, the bending vibration, ν_3, as B_{2g} and the ring deformation, ν_4, as B_{2u}. The C_2' and C_2'' axes could alternatively be defined as bisecting the F–Xe–F angles: then ν_5, ν_3 and ν_4 are described as B_{2g}, B_{1g} and B_{1u} respectively. Some confusion exists in the literature on this point and the notation given in Fig. 14.1 is that recommended by Hagen and other authors.

Assignments made to bands observed[13] in the Raman and infrared spectra are given in Table II. For D_{4h} symmetry, all the infrared-active transitions are found in the ungerade (subscript u) symmetry classes: all the Raman-active transitions belong to the gerade

TABLE II

Normal Vibrations* of the XeF_4 Molecule (D_{4h} symmetry)

Class	Activity	Vibration	Frequency (cm^{-1})
A_{1g}	R	ν_1	543
A_{2u}	i.r.	ν_2	291
B_{2g}	R	ν_3	235
B_{2u}	Inactive	ν_4	
B_{1g}	R	ν_5	502
E_u	i.r.	ν_6	586
E_u	i.r.	ν_7	—

* Notation according to G. Hagen, *Acta Chem. Scand.* **21**, 465 (1967).
R = Raman active; i.r. = infrared active.

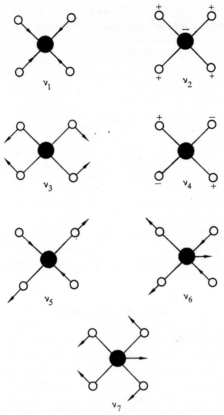

FIG. 14.1. Fundamental vibrations of the planar XeF_4 molecule.
$+$ and $-$ indicate movements of atoms above and below the molecular plane.

(subscript g) symmetry classes. This is because a square planar molecule is centrosymmetric and the spectroscopic mutual exclusion rule is obeyed. In the case of XeF_4, the bands observed in the infrared are not found in the Raman spectrum and *vice versa* and this is strong experimental evidence to support the planar structure of XeF_4.

As a help to the correct assignment of vibrational bands, their characteristic contours provide useful information. The infrared bands observed on a gaseous sample show rotational fine structure. This is generally not resolved but the band contours are valuable pointers to the type of vibration which causes them.

A square planar molecule like XeF_4 is described by three moments of inertia, I_x, I_y and I_z, about three mutually perpendicular axes, x, y and z. The z-axis is the principal symmetry axis, perpendicular to the plane of the molecule and passing through its centre. The x- and y-axes are mutually at right angles in the plane of the molecule. $I_z > I_y = I_x$ and the molecule behaves as an oblate symmetric top.

The two quantum numbers J and K are used to describe transitions between rotational levels. J is a rotational quantum number which governs the total angular momentum and K defines the momentum about the x- or y-axis. Infrared-active vibrations of a symmetric top are classified as parallel or perpendicular depending on whether the oscillating dipole moment associated with the vibration is parallel or perpendicular to the principal symmetry axis.

140

For an oscillating moment parallel to the z-axis, for example, ν_2, which involves out-of-plane motion of the atoms, an absorption band with three branches, P, Q and R results. P, Q and R refer respectively to the ΔJ values of -1, 0 and $+1$ allowed by selection rules for rotational transitions ($\Delta K = 0$ in each case). The Q branch is expected to be very prominent compared with the other two. For an oscillating moment perpendicular to the z-axis, such as ν_6 or ν_7, P, Q and R branches are again expected ($\Delta J = 0$ and ± 1 as before but now $\Delta K = \pm 1$) but the intensities of all three should be comparable.

The contour of the band located at 291 cm^{-1} shows a very intense Q branch so the assignment of this to ν_2 is quite certain.

More recent measurements have confirmed earlier assignments for ν_1, ν_2, ν_3, ν_5, and ν_6. ν_7 and ν_4 are not observed directly in the spectra and have been calculated[14] indirectly using experimental data from studies of the equilibria in the xenon–fluorine system in the temperature range 250–500°C. From the temperature variation of the equilibrium constant for the reaction $XeF_2 + F_2 \rightleftharpoons XeF_4$, values for ΔG^0, ΔH^0 and ΔS^0 were calculated. Theoretical values of ΔS^0 for this reaction at different temperatures were calculated from molecular parameters such as the bond length, established by electron diffraction, the experimental values of ν_1, ν_2, ν_3, ν_5 and ν_6 and estimated values of ν_4 and ν_7. The best values for ν_4 and ν_7 were taken to be those for which the weighted sum ΔS^0(exptl.) − ΔS^0(theor.) was zero. In this way, the values calculated were $\nu_4 = 237$ and $\nu_7 = 250$ cm^{-1}.

The agreement between theoretical and calorimetric standard entropy values[15] is not completely satisfactory. The standard entropy of gaseous XeF_4 at 25°C has been calculated from molecular data as $316 \cdot 6 \pm 1 \cdot 5$ J deg^{-1} mol^{-1}. The value found from calorimetric data is 305 ± 10 J deg^{-1} mol^{-1}. The difference between the two is barely within experimental error and indicates the need for more reliable thermodynamic and molecular data for XeF_4.

Mössbauer Spectra

The Mössbauer effect has been observed with two isotopes of xenon, ^{129}Xe and ^{131}Xe. The abundances of these are 26·44 and 12·18% respectively. The first excited state of ^{129}Xe, formed by β-decay of ^{129}I, is generally used as the source of γ-rays. This excited state, for which $I = 3/2$, loses γ-radiation (6·4 fJ) in the transition to the ground state,

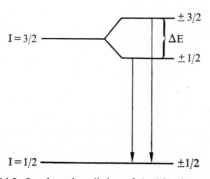

FIG. 14.2. Quadrupole splitting of the Mössbauer spectrum of XeF$_4$.

for which $I = \frac{1}{2}$. Provided some fraction of the γ-rays is emitted without nuclear recoil, resonance absorption of this can occur by xenon nuclei in the ground state and a Mössbauer spectrum is observed. The magnitude of the fraction depends, *inter alia*, on the energy of the γ-radiation and the temperature. It decreases with the exponential square of the energy and increases as the temperature decreases. This relationship with energy renders ^{131}Xe (energy of γ-radiation $= 12\cdot8$ fJ) less suitable than ^{129}Xe for Mössbauer spectroscopy. Even with ^{129}Xe, it is necessary to make measurements at liquid helium temperature $-269°C$ to observe a large enough effect.

The excited, but not the ground, state of ^{129}Xe has a nuclear quadrupole moment, Q, and the interaction between this and the gradient of the electric field at the nucleus results in a quadrupole splitting of the Mössbauer spectrum, Fig. 14.2. In the uncombined xenon atom, the filled shells of electrons have spherical symmetry, they do not contribute to the field gradient at the nucleus and the excited energy level remains degenerate. In square planar XeF_4, the field gradient at the xenon nucleus has symmetry around the z-axis and the excited energy level divides into two states, separated by the nuclear quadrupole splitting,

$$\Delta E = \frac{e^2 qQ}{2}$$

where

$$eq = \frac{\delta^2 \phi}{\delta z^2}$$

the field gradient along the z-axis and $\phi =$ the potential. The velocity spectrum of XeF_4[16], which illustrates this quadrupole splitting, is shown in Fig. 14.3.

Assuming that the bonding in XeF_4 is mainly associated with the filled $5p_x$ and $5p_y$ orbitals of xenon and that the charge distribution is stretched out so much that the contribution of the four bonding electrons to the field gradient effectively vanishes, the field gradient at the nucleus may be regarded as due to the p_z electrons only. Using this model, the quadrupole splitting has been estimated as $54\cdot2$ mm sec^{-1} compared with the experimental value of 42 mm sec^{-1}. The difference is large enough to indicate that the above description of the bonding is only an approximation.

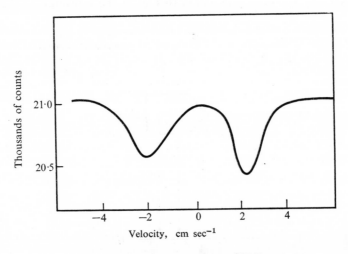

FIG. 14.3. Mössbauer spectrum of XeF_4.

A knowledge of the Mössbauer spectrum of XeF_4 has made it possible to demonstrate the formation of xenon tetrachloride, $XeCl_4$, by β-decay from $K^{129}ICl_4 \cdot H_2O^{17}$. $XeCl_4$ was identified by its Mössbauer spectrum which shows the quadrupole splitting observed with XeF_4.

Theories of Bonding

The existence of fluorides of xenon and other compounds of the noble gases has presented theoretical chemists with some intriguing problems.

Firstly, it is necessary to explain why these compounds should be formed at all. This is specially important in view of the assumption made for so long about the chemical inertness of the noble gases. A significant experimental fact in this context is that those compounds which appear to be the most readily formed are between very electronegative ligands and a heavy central atom. Secondly, an adequate theory should rationalize the stoichiometries and shapes of the molecules. Thus xenon forms binary compounds containing an even number of fluorine atoms but none are known with an odd number. Again, XeF_2 is linear, not triangular, and XeF_4 is square planar, not tetrahedral. Thirdly, a real test of the validity of any theory is the prediction of the formation and properties of other compounds prior to their discovery.

Coulson[18] has discussed the four theoretical models of bonding in xenon fluorides which have had some prominence in the literature. These are (1) the hybridization model, (2) the electron correlation model, (3) molecular orbital theory, and (4) valence-bond resonance theory.

(1) *The hybridization model.* The description of the bonds in terms of electron pairs is preserved if sp^3d^2 hybrids are constructed from the atomic orbitals of xenon. The xenon atom provides eight valence electrons and each of the fluorines provides one in the formation of six electron pairs (four bonding and two lone pairs). The lone pairs are above and below the molecular plane (Fig. 14.4) and directed away from each other in accordance with the Sidgwick–Powell concept that lone pairs on an atom in a molecule are as far apart as possible to minimize repulsion between them. The four remaining pairs are bonding and co-planar. This accounts for the shape of XeF_4 and a similar argument correctly leads to a linear molecule for XeF_2.

This theory is useful because it describes molecular shapes in a simple way but is unsatisfactory if quantitative aspects of the bonding are considered. The excitation energy for the promotion $5s^25p^6 \rightarrow 5s^25p^55d$ is of the order of 1·6 aJ. Two such promotions

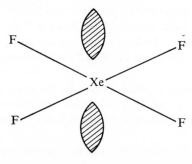

Fig. 14.4. The square planar shape of XeF_4 interpreted according to the Sidgwick–Powell concept.

are necessary if sp^3d^2 hybrids are to be formed and the total energy required is much larger than that expected to be released in the formation of XeF_4. From energy considerations, it therefore seems unlikely that a stable compound could be formed in this way.

This model does not account for the existence of XeF_6 because this molecule requires fourteen valence electrons which cannot be accommodated in octahedral hybrids. These considerations led Coulson to the conclusion that the hybridization model for xenon fluorides is inadequate.

(2) *The electron correlation model.* In the free xenon atom of valence-shell configuration s^2p^6, we may regard each p-orbital as occupied by one electron with α-spin and another with β-spin. Because of their charge, two electrons in a given orbital keep apart from each other but because of their opposite spins, they tend to be rather closer together than if their spins were parallel.

Although the charge associated with the six $5p$-electrons is distributed in a spherically symmetric way, repulsion between them results, according to this model, in an octahedral arrangement on the surface of a sphere having the atomic radius. Four electrons, of alternate spin α and β, are equidistant from each other on the equator and one pole has an electron of spin α, the other of spin β. Bonding with fluorine then occurs in the following way. A fluorine atom with an unpaired electron of spin α approaches the xenon atom with this arrangement of electrons and is attracted to the xenon orbital occupied by an electron of β-spin and is repelled by the orbital occupied by an electron of α-spin. Bonding results from the overlap between xenon and fluorine orbitals containing electrons of opposite spin. Two fluorines can be bound to give a linear molecule and four to give a square-planar one. For each pair of $5p$-electrons decoupled on the formation of two Xe–F bonds, the contribution of the correlation energy has been estimated as 0·08 to 0·16 aJ per bond.

The disadvantages of this model have been discussed.[18,19] The symbols α and β refer to components of the spin and not to the spin itself and there is no clear physical interpretation of these. The correlation energy appears to be independent of atomic size and the model does not explain the importance of this and the electronegativity of the ligand atoms in determining whether or not a noble gas compound is formed. Coulson has concluded that correlation effects 'in so far as they exist at all, are unlikely to provide significant bonding'.

(3) *Molecular orbital theory.* According to this, the bonding is chiefly due to overlap between four $2p_\sigma$-type orbitals on the fluorines and two $5p_\sigma$-type orbitals on the xenon. Molecular orbitals formed in XeF_2 from two fluorine orbitals and one xenon orbital are illustrated in Fig. 14.5. 1σ is a filled bonding orbital, 2σ is a filled non-bonding orbital (confined to the fluorine atoms) and 3σ is an empty anti-bonding orbital. Contributions to the σ-bonding from s or d xenon orbitals are regarded as not significant.

The grouping $1\sigma^2 2\sigma^2$ has been described by Rundle[20] as a three-centre four-electron bond. Its formation necessarily withdraws electron density from xenon and places it on fluorine so the Xe,F bonds have some degree of ionic character, and may be represented by $F^{\delta-}Xe^{2\delta+}F^{\delta-}$.

There will also be π-molecular orbitals formed by the interaction of p-orbitals of appropriate symmetry on the xenon and fluorine atoms. The π-orbitals are doubly degenerate and bonding and anti-bonding orbitals are filled equally so there is no net contribution to the bonding.

In XeF_4, the four fluorine orbitals and the two xenon orbitals combine to form two three-centre four-electron bonds. There is experimental support for the involvement of two of the $5p$-orbitals on xenon from the Mössbauer spectrum described earlier.

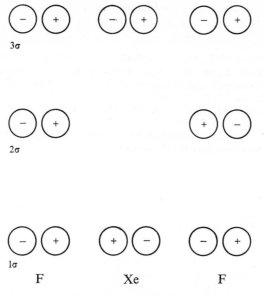

FIG. 14.5. Molecular orbitals in XeF_2.

The concept of a three-electron bond accounts satisfactorily for the planarity of the molecule and emphasizes the importance of the high electronegativity of the ligand atoms in the formation of stable noble gas compounds.

(4) *Valence-bond resonance theory.* In terms of Pauling resonance structures, important contributing forms to the bonding in xenon difluoride are: $FXeF$; $F–Xe^+F^-$ and $F^-Xe^+–F$; F^+XeF^- and F^-XeF^+; and $F^-Xe^{2+}F^-$. The relative contribution of each form has been calculated and the resultant charge distribution found to approximate to $F^{-\frac{1}{2}}Xe^{+1}F^{\frac{1}{2}-}$. The bonds in XeF_4 are also similarly semi-ionic in character. An essential feature of this and the molecular orbital theory is the transfer of electronic charge from xenon to fluorine.

Coulson has concluded that the molecular orbital or the valence-bond resonance theory should preferably be used to describe the bonding in noble-gas compounds. The bonds are essentially no different from those found in other chemical species such as the polyhalide ions and it is not necessary to postulate a new kind of chemical bond to account for the formation of noble-gas compounds.

References

1. YOST, D. M., and KAYE, A. L., *J. Amer. Chem. Soc.* **55**, 3890 (1933).
2. BARTLETT, N., *Proc. Chem. Soc.* 218 (1962).
3. CLAASSEN, H. H., SELIG, H., and MALM, J. G., *J. Amer. Chem. Soc.* **84**, 3593 (1962).
4. HOLLOWAY, J. H., and PEACOCK, R. D., *Proc. Chem. Soc.* 389 (1962).
5. KIRSCHENBAUM, A. D., STRENG, L. V., STRENG, A. G., and GROSSE, A. V., *J. Amer. Chem. Soc.* **85**, 360 (1963).
6. *Inorganic Syntheses*, vol. VIII, p. 254 (1966).
7. BARTLETT, N., and STADKY, F. O., *J. Amer. Chem. Soc.* **90**, 5316 (1968).
8. SIEGEL, S., and GEBERT, E., *J. Amer. Chem. Soc.* **85**, 240 (1963).
9. TEMPLETON, D. H., ZALKIN, A., FORRESTER, J. D., and WILLIAMSON, S. M., *J. Amer. Chem. Soc.* **85**, 242 (1963).

10. IBERS, J. A., and HAMILTON, W. C., *Science* **139**, 106 (1963).
11. BURNS, J. H., AGRON, P. A., and LEVY, H. A., *Science* **139**, 1209 (1963).
12. BOHN, R. K., KATADA, K., MARTINEZ, J. V., and BAUER, S. H., *Noble Gas Compounds*, University of Chicago Press, p. 238 (1963).
13. CLAASSEN, H. H., CHERNICK, C. L., and MALM, J. G., *J. Amer. Chem. Soc.* **85**, 1927 (1963).
14. WEINSTOCK, B., WEAVER, E. E., and KNOP, C. P., *Inorg. Chem.* **5**, 2189 (1966).
15. SCHREINER, F., McDONALD, G. N., and CHERNICK, C. L., *J. Phys. Chem.* **72**, 1162 (1968).
16. PERLOW, G. J., and PERLOW, M. R., *Rev. Mod. Phys.* **36**, 353 (1964).
17. PERLOW, G. J., and PERLOW, M. R., *J. Chem. Phys.* **41**, 1157 (1964).
18. COULSON, C. A., *J. Chem. Soc.* 1442 (1964).
19. MALM, J. G., SELIG, H., JORTNER, J., and RICE, S. A., *Chem. Rev.* **65**, 199 (1965).
20. RUNDLE, R. E., *J. Amer. Chem. Soc.* **85**, 112 (1963).

CHAPTER 15

CHROMIUM(II) ACETATE

Introduction

In the first transition series, the metals titanium, vanadium and chromium are most stable in their higher oxidation states and states below +3 are not stable in aqueous solution unless oxygen is completely absent. The strong reducing power of chromium(II) is evident from the standard reduction potential, E^0, for the half-reaction:

$$Cr^{3+} + e \rightleftharpoons Cr^{2+} \quad E^0 = -0.41 \text{ V}$$

Chromium(II) acetate is one of a number of compounds which are, however, relatively stable. Although precautions must be taken to prevent access of oxygen during its preparation, the compound is almost completely insoluble in water and stable in air. The reason for these properties is directly concerned with the structure of and the kind of bonding within the solid. The acetate is dimeric and the dimer is characterized by the existence of bonds directly between metal atoms and the bridging action of acetate groups.

As well as these features, which are of interest to inorganic chemists, chromium(II) acetate is used to an increasing extent as a reducing agent in organic chemistry.

Preparation and Reactions

The preparation of chromium(II) acetate is conveniently carried out in a nitrogen atmosphere by the reduction of chromium(III) chloride solution with zinc amalgam,[1] followed by the addition of aqueous sodium acetate to precipitate the dihydrate of the chromium(II) salt as a deep red solid. The monohydrate is obtained by drying the dihydrate *in vacuo* for several hours at room temperature. Complete dehydration is achieved by drying *in vacuo* at 100°C. The anhydrous compound is light brown in colour and even more susceptible to oxidation than either hydrate.

Chromium(II) acetate has also been made by dissolving electrolytic (>99·99% pure) chromium in acetic acid in the absence of oxygen.[2] This method gives a product free from contamination by the chromium(III) salt.

The acetate is a useful starting material for the preparation of other chromium compounds. The bis(acetylacetonate) of chromium(II) has been made[3] by heating the acetate and acetylacetone together in a nitrogen atmosphere in a sealed tube. The acetate reacts with 2-2'-dipyridyl in aqueous suspension with disproportionation[4] and part of the chromium is converted to tris(2-2'-dipyridyl) chromium(0) and the rest to chromium(II) and (III) complexes. In tetrahydrofuran, the reaction with 2-2'-dipyridyl gives bis(dipyridyl) chromium(II) acetate.[5]

Structure

Chromium(II) acetate forms monoclinic crystals belonging to the space group $C2/c$. The unit cell contains[6] four $Cr_2(CH_3COO)_42H_2O$ molecules. In this dimer, the two chromiums are joined by four bridging acetato-groups in such a way that the four oxygens, arranged approximately at the corners of a square, are the nearest neighbours of each metal atom (Fig. 15.1). The bond length, Cr,O = 0·197 nm. The octahedral coordination of chromium

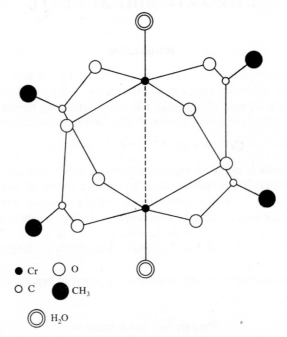

FIG. 15.1. The structure of dimeric chromium(II) acetate dihydrate.

is completed by a water molecule, $Cr,O(H_2)$ = 0·220 nm, above the plane of the square and the second chromium below, Cr,Cr = 0·264 nm. The main intermolecular forces are hydrogen bonds and each dimeric molecule is linked by eight such bonds to four neighbouring molecules. This bonding accounts for the low solubility in water of the acetate compared with other chromium(II) compounds, many of which, like the perchlorate and chloride, are very soluble.

The Cr,Cr distance in the dimeric molecule is only slightly longer than the shortest interatomic distance (0·2498 nm) observed in metallic chromium at ordinary temperatures and it must be concluded that there is direct bonding between the two metal atoms of the dimer. Other transition metal compounds are known in which similar metal–metal bonding is indicated by the unusual shortness of the intermetallic distance as revealed by X-ray

148

analysis. These include the cluster compounds exemplified by molybdenum dichloride and the polynuclear iron carbonyls which are described elsewhere (p. 167).

Another diagnostic feature of metal–metal bonds is the abnormally small or zero magnetic moment of compounds containing these. Chromium(II) acetate monohydrate shows a very slight paramagnetism.[7] Cr^{2+} is a d^4 ion and if we consider only the two electron-pair bonds formed between chromium and two acetate groups, this leaves two unpaired electrons on the metal which should confer marked paramagnetism on the compound. On the assumption that the observed weak paramagnetism is probably due to impurities such as chromium(II) not combined in dimeric form or chromium(III), the pure acetate may be regarded as diamagnetic. The unpaired d-electrons on each chromium are paired up with those on another to form metal–metal bonds.

A comparable situation is found in the case of copper(II) acetate. The molecular unit[8] in the solid monohydrate is again a dimer, the Cu,Cu distance therein (0·264 nm) is only slightly greater than that (0·256 nm) found in metallic copper, and the magnetic moment is also unusually low. The structure of copper(II) acetate has been described in terms of direct bonding between metal atoms. This would account for the magnetic properties, although an alternative explanation involving a super-exchange mechanism via the acetate ligands cannot be excluded.

References

1. *Inorganic Syntheses*, vol. VIII, p. 126.
2. Lux, H., and Illmann, G., *Chem. Ber.* **91**, 2143 (1958).
3. Costa, G., and Puxeddu, A., *J. Inorg. Nucl. Chem.* **8**, 104 (1958).
4. Herzog, S., and Renner, K. C., *Chem. Ber.* **92**, 872 (1959).
5. Herzog, S., Oberender, H., and Paul, S., *Z. Naturforsch.* **18B**, 158 (1963).
6. Van Niekerk, J. N., Schoening, F. R. L., and deWet, J. F., *Acta Cryst.* **6**, 501 (1953).
7. Furlani, C., *Gazz. Chim. Ital.* **87**, 876 (1957).
8. Van Niekerk, J. N., and Schoening, F. R. L., *Nature* **171**, 36 (1953).

CHAPTER 16

DIBENZENE CHROMIUM

Introduction

Remarkable progress in organometallic chemistry followed the discovery of ferrocene in 1951. A wide range of cyclopentadienyl complexes was soon synthesized and the aromatic character generally shown by cyclopentadienyl rings bound to a transition metal was established. A logical extension of this work was the investigation of the possibility of preparing metal complexes containing other conjugated ring systems and, in 1955, the first example of organometallic complexes derived from benzene was synthesized.[1] This was bis(benzene) chromium(0), a sandwich molecule in which the chromium atoms lie between the two benzene rings.

A number of other bis(benzene) metals is now known but this is comparatively small in relation to the bis(cyclopentadienyl) complexes which have been made. Benzene complexes are generally air-sensitive and quite easily decomposed. A study of their properties is therefore associated with some experimental difficulty.

Preparation

Bis(benzene) chromium(0) is made via the bis(benzene) chromium(I) cation. This is formed in the reduction of chromium(III) chloride by aluminium in the presence of aluminium chloride as catalyst and of benzene:[2]

$$3CrCl_3 + 2Al + AlCl_3 + 6C_6H_6 \rightarrow 3[Cr(C_6H_6)_2][AlCl_4]$$

The reaction must be performed with thoroughly dry reactants and in the absence of oxygen. The complex tetrachloro–aluminate is treated successively with methanol and water to cause hydrolysis and to produce a solution containing the yellow ion, $Cr(C_6H_6)^+$. This is reduced to $Cr(C_6H_6)_2$ by sodium dithionite. Bis(benzene) chromium(0) can be extracted by benzene and, after removal of the solvent, obtained as dark brown crystals.

Pure bis(benzene) chromium melts at 284–285°C. At about 300°, rapid decomposition occurs to metallic chromium and benzene. The compound is soluble in benzene and toluene, but only sparingly soluble in solvents like ether, alcohol and petroleum ether.

150

Reactions

Salts containing $Cr(C_6H_6)_2{}^+$ are made by the aerial oxidation of $Cr(C_6H_6)_2$. In the presence of anions like reineckate, picrate, tetraphenylborate, perchlorate or iodide, a sparingly soluble salt is produced. More directly, these salts are obtained by precipitation from the solution obtained by hydrolysis of $[Cr(C_6H_6)_2][AlCl_4]$. These compounds are generally light-sensitive and are rapidly decomposed by aqueous acid.

It is difficult to prepare derivatives of bis(benzene) chromium(0) by direct substitution because of the ease with which the complex breaks down or is oxidized. However, it seems clear that the benzene rings do not show the reactivity which might have been anticipated from the reactions of the cyclopentadienyl rings in ferrocene.

Attempted Friedel–Crafts reactions with acetyl and benzoyl chlorides in the presence of aluminium chloride leads[3] to a breakdown of the complex and the formation of the corresponding organic derivative. Mercuration of the organic rings, which has been effected in the case of ferrocene, cannot be achieved because of the oxidizing action of mercury(II) salts which produces $Cr(C_6H_6)_2{}^+$. Attempted nitration and sulphonation also results in oxidation. No reaction is observed with sodamide. These negative results indicate that the organic rings in bis(benzene) chromium(0), in contrast to those in ferrocene, no longer have the capacity to undergo substitution reactions characteristic of aromatic compounds.

Preparative methods for substituted derivatives of bis(benzene) chromium have been developed[4] by exploiting the metallation of the benzene rings using lithium or sodium alkyls. For example, the reaction between bis(benzene) chromium and amyl sodium gives a metallated derivative which reacts with carbon dioxide to form the sodium salts of mono-, di- and poly-substituted bis(benzene) chromium carboxylic acids:

$$Cr(C_6H_6)_2 \xrightarrow{NaC_5H_{11}} Cr(C_6H_5Na)_2 \xrightarrow{CO_2} Cr(C_6H_5COONa)_2$$

These may be separated by esterification using dimethyl sulphate, followed by chromatography of the ester mixture. The dimethyl ester of the dicarboxylic acid isolated from the mixture has been shown to consist of a mixture of the isomers which are possible when both substituents are in the same ring, the meta isomer being the main component. This behaviour contrasts with that of the bis(cyclopentadienyl)s of iron, ruthenium and osmium which, on substitution, give monocarboxylic acids and exclusively those dicarboxylic acids wherein each ring carries one substituent.

Other substituted bis(benzene) chromium complexes have been prepared[5] by reaction of its disodium derivative variously with formaldehyde, acetaldehyde, benzaldehyde and diphenylketone. In each case, monosubstituted complexes can be isolated from the mixture of reaction products. The constitution of these complexes has been determined by thermal decomposition followed by gas chromatographic analysis of the arenes formed. Thus, when bis(benzene) chromium is metallated with amyl chloride and sodium and then treated with paraformaldehyde and water, one of the products is benzylalcohol benzene chromium(0).

Reactions between alkyl halides, RX, and bis(benzene) chromium at room or elevated temperature or under the influence of ultraviolet light proceed[6] largely with oxidation to $(C_6H_6)_2CrX$ and hydrocarbons. Chromium, benzene and hydrogen have all been identified as by-products. Deuteration studies have indicated that part of the hydrogen comes from the benzene rings bonded to chromium, presumably being released when the alkyl groups substitute in the rings.

Bis(benzene) chromium reacts with Lewis acids such as $1:3:5$ trinitrobenzene, p-quinone, chloranil and tetracyanoethylene to produce deeply-coloured $1:1$ adducts. Ultraviolet, infrared and E.S.R. spectroscopy have indicated that this type of reaction proceeds with electron transfer from bis(benzene) chromium to the acceptor molecule forming $(C_6H_6)_2Cr^+$ and the corresponding radical anion.

Structure

The 'sandwich' shape of the bis(benzene) chromium molecule and the crystal structure of the solid were established using X-ray diffraction by Weiss and Fischer[7] in 1956. The crystal is cubic with a unit cell containing four molecules of $(C_6H_6)_2Cr$ and having a cell edge of $0·967$ nm. The molecule is centrosymmetric and belongs to the point group, D_{3d}. It can also belong to any group of higher symmetry which has D_{3d} as a sub-group. One possibility is the symmetry group D_{6h} in which the benzene rings are parallel and eclipsed. This model has a six-fold symmetry axis normal to both rings and this passes through the metal atom. There is also a horizontal plane of symmetry lying at right angles to this axis which bisects the metal atom. For the molecule to possess a six-fold axis all the C–C bond lengths within each ring must be identical. On the assumption of D_{6h} symmetry, Weiss and Fischer calculated the following values for the bond lengths: $C,C = 0·138 \pm 0·005$; $Cr,C = 0·219 \pm 0·01$ nm.

There has been some dispute over whether bis(benzene) chromium has a six-fold symmetry or not. Thus, Jellinek, on the basis of his X-ray diffraction data on a crystal at room temperature,[8] has concluded that the molecule has D_{3d} symmetry because of an alternation of C,C bond lengths within the benzene rings, the calculated values for adjacent bonds being $0·1436 \pm 0·0012$ and $0·1366 \pm 0·0012$ nm.[9] The standard deviation of these values is much less than the difference between them so it appears that the alternation is statistically significant. Jellinek further concluded that the ligand rings are slightly, but not very significantly, puckered.

On the other hand, Cotton, Dollase and Wood[10] carried out an independent X-ray analysis and decided there was no significant deviation from D_{6h} symmetry. These investigators reported an average C,C bond length of $0·1387$ nm and that alternate bonds differed by less than $0·002$ nm. Some slight degradation of the six-fold symmetry is believed to occur due to the effect of the crystalline environment rather than to any significant difference in the C,C bond lengths. Ibers[11] has analysed the same experimental data and confirmed their consistency with D_{6h} symmetry.

Some kind of orientational disorder in the crystal state could account for the different conclusions reached by various investigators. However, this seems unlikely in view of the X-ray diffraction data obtained on a sample at a low temperature 173°C, where the thermal motions of the atoms are very small and where their positions can be very accurately determined. These confirm D_{6h} symmetry and the C,C bond length is $0·1419$ nm.

Electron diffraction measurements on gaseous bis(benzene) chromium have shown[12] that the molecule has D_{6h} symmetry in the vapour phase. The bond lengths found are: $C,C = 0·1423 \pm 0·0002$; $C,H = 0·1090 \pm 0·0005$; Cr to C distance $= 0·2150 \pm 0·0002$; vertical ring-to-ring distance $= 0·3226 \pm 0·0005$ nm.

Two mono-arene carbonyl complexes related to bis(benzene) chromium are hexamethyl-benzene chromium tricarbonyl and benzene chromium tricarbonyl. The X-ray analysis of the first of these has established[13] that the carbon atoms of the methyl groups and those in the benzene ring are co-planar and that there is no distortion of the benzene ring such as the alternation of C,C bond lengths. Similarly, it has been found that the benzene ring in benzene chromium tricarbonyl retains[14] the D_{6h} symmetry of free benzene. This symmetry appears to be found whether the chromium complex is of the mono- or the bis-arene variety.

Vibrational Spectrum

Conclusions about the molecular symmetry of bis(benzene) chromium have also been reached from measurements of its vibrational spectrum.

In complexes between cyclic systems, C_nH_n, and metals it has been found useful to make a formal distinction between vibrations within the ligands and those of the molecular framework or 'skeleton' of the whole complex, assuming for this purpose that the ligand rings are rigid. For a sandwich molecule like bis(benzene) chromium, two extreme situations can be defined in which there are respectively strong and little or no coupling through the central atom between the vibrations of the two ligands.[15,16] In the latter case, the vibrational spectrum can be treated as though made up of the ligand spectrum and that of the ligand–metal–ligand framework, regarded as a quasi-three-mass model.

Benzene itself belongs to the symmetry group D_{6h} and its vibrational spectrum[17] has been interpreted accordingly. In the bis(benzene) chromium complex, the symmetry of the rings must be considered in relation to the symmetry of the whole molecule. If there is little or no coupling, the symmetry of each ring may be regarded as lowered to C_{6v} local symmetry by bonding to the metal. If there is strong coupling, then the symmetry must be defined for the whole molecule. As this is known, from X-ray diffraction work, to possess a centre of inversion, the most likely symmetry is D_{6h} or D_{3d}. Other possibilities for a sandwich molecule, such as D_{3h} and C_{3v}, cannot apply in this case because they do not possess a centre of inversion. Three possible structural models[15] are shown in Fig. 16.1. In (a) the rings are planar and the C,C bonds are of equal length, the complex as well as the ligands possessing D_{6h} symmetry. In (b) the rings are puckered and the C,C bonds are equal. Both the complex and the ligands have D_{3d} symmetry. In (c) the rings are planar but the C,C bonds are alternately short and long. The complex belongs to the point group D_{3d} and the ligand to D_{3h}.

The number of ligand vibrations expected for different models of the bis(benzene) chromium molecule is shown in Table I. This indicates that the same number of frequencies in both the infrared and the Raman spectra is predicted for the six-fold symmetry groups,

TABLE I

Ligand Vibrations for Different Symmetries of the bis(benzene) Chromium Molecule

	Symmetries		
	Strong coupling		Weak coupling
	D_{3d}	D_{6h}	C_{6v}
Number of vibrations active in infrared	17	7	7
Number active in Raman	17	13	13
Number completely inactive	6	20	7

(a)

(b)

(c)

FIG. 16.1. Possible structures of bis(benzene) chromium.

D_{6h} and D_{6v}. It is therefore not possible to distinguish between models in which there is weak and strong coupling on the basis of the vibrational spectrum.

However, the symmetries D_{6h} and D_{3d} may be distinguished by the total number of active normal vibrations and the number of those vibrations which appear only in the Raman spectrum. Because there is a centre of inversion, the mutual exclusion rule must apply to the vibrations of the complex. If there is only small coupling between the in- and out-of-phase vibrational modes of the benzene rings, 'near-coincidences' would be observed in the frequencies of lines in the Raman and infrared spectra. Seven such coincidences would be observable for D_{6h} symmetry and seventeen for D_{3d}.

The framework or 'skeletal' vibrations are the same in number and type whether the symmetry is D_{6h} or D_{3d}. These cannot therefore help in differentiating the two symmetries. Skeletal vibrations for D_{6h} symmetry are shown schematically in Fig. 16.2; three are infrared active, two are Raman active and one is totally inactive.

Detailed Raman and infrared spectroscopic measurements have been made[15,16] on compounds containing $Cr(C_6H_6)_2{}^+$. Infrared results are also available for the uncharged complex but Raman work is hampered by its dark colour and low solubility in solvents except those which are benzene-like in character. For these reasons, only the skeletal vibrations have been observed for benzene solutions of $Cr(C_6H_6)_2$.

For $Cr(C_6H_6)_2$ I, the number of near-coincidences and of vibrations appearing only in the Raman spectrum have confirmed that the complex ion has D_{6h} symmetry. A detailed analysis of the spectrum shows best agreement with predictions for D_{6h} symmetry of $Cr(C_6H_6)_2$ itself and for C_{6v} symmetry of the benzene molecules.[18,19]

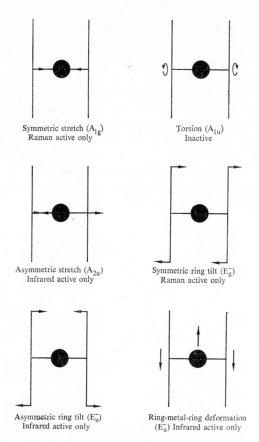

FIG. 16.2. The skeletal vibrations of a bis(benzene) metal complex of D_{6h} symmetry.

Skeletal vibrations have been identified with bands observed in the spectrum of $Cr(C_6H_6)_2$ at $270(A_{1g})$, $459(A_{2u})$, $332(E_g^-)$, $490(E_u^-)$ and $140(E_u^-)$ cm^{-1}. Corresponding bands are observed at slightly different frequencies in $Cr(C_6H_6)_2$ I. A_{1u}, not directly observable, has been estimated as 305 cm^{-1} from combination bands.

Excluding the two C–H stretching modes (which should occur around 3000 cm^{-1}), the other five infrared active fundamentals of the ligand should occur in the region 1600 to 500 cm^{-1} if only moderate shifts in frequency occur on going from benzene itself to $Cr(C_6H_6)_2$. Only one band, located at 3047 cm^{-1}, has been found[18] for the C–H mode in the infrared spectrum of crystalline bis(benzene) chromium at $-180°C$. Five strong bands, at 1430, 1014, 1002, 970 and 796 cm^{-1}, have been related to bands found at 1478, 1036, 990, 854 and 687 cm^{-1} in the spectrum of benzene vapour. Fritz[16] has assigned a band found at 833 cm^{-1} as a fundamental instead of that at 1014 cm^{-1}.

The symmetry of $Cr(C_6H_6)_2$ in the vapour phase has been established[19] as D_{6h} and eight out of the nine infrared active modes expected in the region 400 to 4000 cm^{-1} have been observed (the total of ten infrared active bands for the whole complex is completed by the E_u^- band observed outside this range, at 140 cm^{-1}). Some additional frequencies are found in the infrared spectrum of the complex in the solid state but these are probably due to the influence of crystal field forces on the spectrum of the ligand.

Electronic Absorption Spectrum

The absorption spectrum of bis(benzene) chromium(I) iodide has been measured in solution and in the crystalline state[20] to determine how the absorption bands of benzene are affected by bonding to the metal. The benzene band located at about 260 nm shows characteristic fine structure. In the absorption spectrum of $[Cr(C_6H_6)_2]$ I in aqueous solution (Fig. 16.3), four significant bands (I to IV) with maxima respectively at 1111, 400, 340·9 and 277·5 nm are observed. I, in the near infrared, and II, the inflection in the visible region, have been ascribed to electronic transitions between energy levels of chromium(I) split by the ligand field. Bands I and II are characteristic of the metal in its normal coordination compounds and this suggests the bonding in $[Cr(C_6H_6)_2]$ I may be similar, that is, involving the withdrawal, to a considerable extent, of the π-electrons on the benzene rings by the central metal atom.

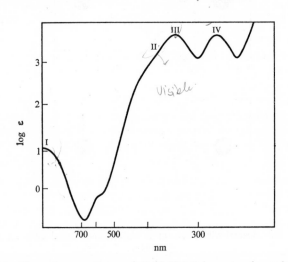

FIG. 16.3. Absorption spectrum of $Cr(C_6H_6)_2$I in water (reproduced with permission from Yamada, S., Yamazaki, H., Nishikawa, H., and Tsuchida, R., *Bull. Chem. Soc. Jap.* **33**, 481 (1960)).

III and IV are much more intense than the ligand-field bands and are probably due to n-π type electronic transitions within the ligands. IV is much more intense than the band found in this region for benzene and does not show any fine structure. Moreover, measurements on benzene in various solvents have shown that the absorption maximum of the band in the near ultraviolet moves to longer wavelengths when the solvent is changed from a less to a more polar one. In contrast, both III and IV show a definite shift to shorter wavelengths when the spectrum of the complex in alcoholic and aqueous solution is compared, the polarity of alcohol being lower than that of water.

An additional intense absorption band found at 225 nm in the spectrum of the iodide in water is a charge-transfer band due to the iodide ion.

The aromaticity of benzene gives rise to marked dichroism,[21] the absorption of light in the direction parallel to the ring being much greater than that normal to the ring. No similar dichroism has been found in $[Cr(C_6H_6)_2]$ I and this confirms the conclusion, from the absorption spectrum in solution, that the arrangement of the electrons in the benzene rings in this complex is very different from that in free benzene.

156

Bonding

Fritz and co-workers[22] have discussed the kind of bonding between the metal and π-bonded aromatic rings like benzene and the cyclopentadienyl anion. Two essentially different concepts have been proposed. These are the pseudo-octahedral and the axially symmetric models. According to the first of these, each ligand is tri-coordinate with respect to the metal. According to the second, the ligand is formally monodentate but different bond orders are possible depending on the electronic energy levels of the molecule and the number of electrons required to fill the available molecular orbitals.

It proves difficult to describe the bonding in all π-bonded metal complexes on the basis of only one of these models.

The properties of transition metal complexes involving benzene can generally be accounted for if benzene is regarded as being tridentate and donating six π-electrons to the metal. The cyclopentadienyl anion is formally described as monodentate, although also a 6π-electron donor. Thus a bis(benzene) metal complex contains a pseudo-octahedral six-coordinated central metal atom and a bis(π-cyclopentadienyl) metal complex a pseudo-linear (resulting from the rotational symmetry around the molecular axis) two coordinated metal atom. The six π-electrons of the ligand can be involved in the bonding in this second type of complex by the filling-up of bonding molecular orbitals. Between the metal and the ligand lies the cloud of bonding electrons which resembles an axially-symmetric cone of symmetry $C_{\infty v}$ and the metal-ligand bond order is greater than one.

The axially symmetric model of symmetry D_{nh}, for π-bonded aromatic complexes of general formula $[(CH)_n]_2M$, approximates, as n increases, to the limiting case of a three-mass, linear molecule of $D_{\infty h}$ symmetry (analogous to CO_2). This model should therefore describe the bis(aromatic ring) metal complex increasingly better as n rises. However, for benzene complexes, the possibilities of overlap between formally mono-, di- and tridentate ligands with the orbitals of the central metal atom favour a tridentate function of the ligand. From considerations of symmetry, the ligand can readily adapt itself to the octahedral coordination of the metal. In this instance, the three-fold symmetry of the benzene ring coincides with the molecular axis of the complex. In contrast, for metal complexes with D_{5h} or D_{7h} symmetry, the three-fold axis of the coordination octahedron coincides with the five- or seven-fold axis of the ring ligands and no octahedral coordination can occur.

This description of bonding in bis(benzene) complexes is supported by observations of the infrared spectrum of $Cr(C_6H_6)_2$ in the crystalline state by Fritz and Fischer.[23] They found a greater number of bands than that expected for D_{6h} symmetry and attributed these to a lowering of molecular symmetry because of distortion of the ligand rings. This is attributed by them partly, but not completely, to the influence of intermolecular forces within the crystal. The more complex spectrum of the complex in the solid state would be consistent with a preferred three-fold symmetry of the benzene ligands about the metal.

In the case of a central atom or ion with a $3d^5$ configuration, as in $Cr(C_6H_6)_2{}^+$ or $V(C_6H_6)_2$, the three-fold symmetry of the ligand is no longer energetically preferred, the complex shows D_{6h} symmetry and can be regarded as a pseudo-linear molecule.

References

1. FISCHER, E. O., and HAFNER, W., *Z. Naturforsch.* **10b**, 665 (1955).
2. FISCHER, E. O., and HAFNER, W., *Zeit. anorg. allgem. Chem.* **286**, 146 (1956).
3. FRITZ, H. P., and FISCHER, E. O., *Z. Naturforsch.* **12b**, 67 (1957).
4. FISCHER, E. O., and BRUNNER, H., *Z. Naturforsch.* **16b**, 406 (1961).

5. FISCHER, E. O., and BRUNNER, H., *Chem. Ber.* **98**, 175 (1965).
6. RAZUVAEV, G. A., DOMRACHEV, G. A., and DRUZHKOV, O. N., *Z. Obshch. Khim.* **33**, 3084 (1963).
7. WEISS, E., and FISCHER, E. O., *Zeit. anorg. allgem. Chem.* **286**, 142 (1956).
8. JELLINEK, F., *Nature* **187**, 871 (1960).
9. JELLINEK, F., *J. Organomet. Chem.* **1**, 43 (1963).
10. COTTON, F. A., DOLLASE, W. A., and WOOD, J. S., *J. Amer. Chem. Soc.* **85**, 1543 (1963).
11. IBERS, J. A., *J. Chem. Phys.* **40**, 3129 (1964).
12. HAALAND, A., *Acta Chem. Scand.* **19**, 41 (1965).
13. BAILEY, M. F., and DAHL, L. F., *Inorg. Chem.* **4**, 1298 (1965).
14. BAILEY, M. F., and DAHL, L. F., *Inorg. Chem.* **4**, 1314 (1965).
15. FRITZ, H. P., LUTTKE, W., STAMMREICH, H., and FORNERIS, R., *Spectrochim. Acta* **17**, 1068 (1961).
16. FRITZ, H. P., *Adv. in Organomet. Chem.* **1**, 240 (1964).
17. MAIR, R. D., and HORNIG, D. F., *J. Chem. Phys.* **17**, 1236 (1949).
18. SNYDER, R., *Spectrochim. Acta* **15**, 807 (1959).
19. NGAI, L. H., STAFFORD, F. E., and SCHAEFER, L., *J. Amer. Chem. Soc.* **91**, 48 (1969).
20. YAMADA, S., YAMAZAKI, H., NISHIKAWA, H., and TSUCHIDA, R., *Bull. Chem. Soc. Jap.* **33**, 481 (1960).
21. YAMADA, S., NAKAMURA, H., and TSUCHIDA, R., *Bull. Chem. Soc. Jap.* **30**, 647 (1957).
22. FRITZ, H. P., KELLER, H. J., and SCHWARZHANS, K. E., *J. Organomet. Chem.* **13**, 505 (1968).
23. FRITZ, H. P., and FISCHER, E. O., *J. Organomet. Chem.* **7**, 121 (1967).

CHAPTER 17

MOLYBDENUM(II) CHLORIDE

Introduction

One of the most intriguing discoveries in inorganic chemistry during the past decade has been the synthesis of some transition metal compounds in which there is clear evidence for the existence of direct bonding between metal atoms. The realization that this type of bonding occurs quite widely has made it possible to understand the properties, hitherto puzzling in some respects, of certain compounds which have been known for many years. For example, the dihalides of molybdenum and tungsten, of empirical formula, MX_2, appear to contain the metal in a low oxidation state but they show little or no reducing power compared with the corresponding chromium(II) halides, which are strong reducing agents. In view of the trend towards more stable higher oxidation states which is normally found with increasing atomic number within a particular sub-group of transition metals, the unexpected stability of these dihalides requires explanation.

They are now recognized as examples of a class of compounds known as metal atom cluster compounds. These have been defined by Cotton[1] as 'those containing a finite group of metal atoms which are held together entirely, mainly, or at least to a significant extent, by bonds directly between metal atoms even though some non-metal atoms may be associated intimately with the cluster'. Thus the 'dichloride' of molybdenum actually contains an octahedral cluster of metal atoms and is better formulated as Mo_6Cl_{12}. Halides of this type are formed principally by the elements Nb, Ta, Mo, W and Re and range from binuclear ions like $Re_2Cl_8^{2-}$, $Tc_2Cl_8^{3-}$ and $W_2Cl_9^{3-}$, through the trinuclear ion, $Re_3Cl_{12}^{3-}$, to compounds like Nb_6Cl_{14}, $NbF_{2.5}$, $TaCl_{2.5}$ and $TaBr_{2.5}$ which all contain octahedral metal atom clusters. A second class of compounds in which metal-to-metal bonds exist is composed of polynuclear carbonyls like $Mn_2(CO)_{10}$, $Fe_2(CO)_9$ and $Fe_3(CO)_{12}$ and related compounds containing nitric oxide or organic π-electron systems.

Following the synthesis and study of the properties of a number of cluster compounds, it is now possible to describe more precisely the factors which determine when metal-to-metal bonds are formed. In the case of a transition metal like molybdenum which shows several oxidation states, it is also instructive to compare the structures and properties of the various binary compounds formed between it and a halogen.

Preparation

Mo_6Cl_{12} can be made by the action of chlorine or phosgene on the heated metal.

The preparation using chlorine is conveniently carried out in a Pyrex tube containing three compartments.[2] Molybdenum powder is placed in two of these. The powder in the first compartment is heated in chlorine gas. $MoCl_5$ is formed and can be condensed before reaching the second compartment. It is then vaporized in a nitrogen gas stream and passed over molybdenum at red heat in the second compartment. This produces $MoCl_3$ which sublimes into the third compartment, where it disproportionates into Mo_6Cl_{12} and $MoCl_5$. Mo_6Cl_{12} is non-volatile and remains behind as the volatile $MoCl_5$ distils off.

The impure Mo_6Cl_{12} can be purified by solution in hot HCl. The complex chloro-acid, $(H_3O)_2[(Mo_6Cl_8)Cl_6]\cdot 6H_2O$, crystallizes out on cooling. When this acid is heated *in vacuo*, pure Mo_6Cl_{12} is obtained.

Mo_6Cl_{12} is a yellow powder which is insoluble in water and organic solvents like toluene and petroleum ether. It shows remarkable stability to oxidation.

Reactions

These are best understood in the light of the observation that one-third of the chlorine atoms in Mo_6Cl_{12} is easily removed. For example, the complex ion, $[(Mo_6Cl_8)Cl_6]^{2-}$, is partially hydrolysed in dilute alkaline solution to give $[(Mo_6Cl_8)(OH)_6]^{2-}$. Further replacement of Cl^- by OH^- occurs slowly in alkaline solution. The reaction obeys second-order kinetics and the rate depends on the concentrations of the chloromolybdate(II) and hydroxyl ions.[3] A crystalline hydroxide, $[(Mo_6Cl_8)(OH)_4]15H_2O$, is formed by the action of dilute aqueous ammonia on the complex chloro-acid.

The structure of Mo_6Cl_{12} (described below) indicates the presence of $Mo_6Cl_8^{4+}$ groupings and its reactions suggest that these are preserved intact when the other four chlorines are replaced by other ligands. It appears that the $Mo_6Cl_8^{4+}$ ion can bind up to six additional donor molecules or ions. Cotton and Curtis[4] have confirmed this by preparing compounds of the types: $[Mo_6Cl_8\cdot L_6](ClO_4)_4$; $[Mo_6Cl_8L_2](CH_3\cdot SO_3)_4$; and $[Mo_6Cl_{12}L_2]$. In each type, L is either dimethylsulphoxide, $(CH_3)_2SO$, or dimethylformamide, $HCON(CH_3)_2$. Infrared spectroscopy has shown that the perchlorate ions are not coordinated to the metal but that $(CH_3)_2SO$, $HCON(CH_3)_2$ and $CH_3SO_3^-$ are (throughtheir oxygen atoms). The bonding of six extra ligands can be understood from the structure of the $Mo_6Cl_8^{4+}$ cluster.

Structure

Mo_6Cl_{12} forms orthorhombic crystals and the unit cell contains four Mo_6Cl_{12} groupings.[5] The structure is composed of $Mo_6Cl_8^{4+}$ clusters, each of which is made up of a cubic arrangement of chlorines (Cl^c) with the molybdenum atoms at the centres of the cube faces. Each molybdenum is linked to four 'cluster' halogens with which it is almost coplanar, to four neighbouring molybdenums and to a fifth chlorine external to the cluster, giving a total coordination number of 9. The Mo,Mo distance in the cluster is 0·261 nm. Figure 17.1 shows the structure of the $Mo_6Cl_8^{4+}$ ion and the location of six chlorines outside this. Four of these, Cl^b, act as bridges between molybdenum atoms of adjacent clusters and the remaining two, Cl^a, are coordinated to only one molybdenum. The different chlorines can

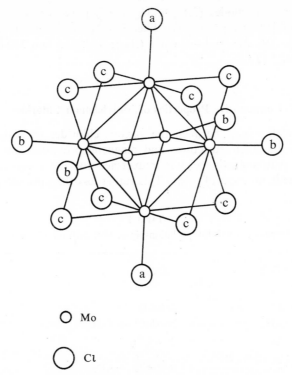

○ Mo

○ Cl

FIG. 17.1. The structure of molybdenum(II) chloride showing the different types (designated a, b and c) of bonded chlorines.

be designated by writing the formula as $\{[Mo_6Cl_8{}^c]Cl_2{}^a\}Cl_{4/2}{}^b$ and the overall stoichiometry is therefore 'MoCl$_2$'.

Vibrational Spectrum

The grouping $[Mo_6Cl_8]Cl_6{}^{2-}$ has intrinsic O_h symmetry although this may be reduced by its environment in crystalline compounds. There are 54 vibrational modes altogether for such a species: $3A_{1g} + 3E_g + 2T_{1g} + 4T_{2g} + A_{2u} + E_u + 5T_{1u} + 3T_{2u}$. Only five of these, of symmetry type T_{1u}, are infrared active and they have been identified by Cotton, Wing and Zimmerman[6] with two Mo–Cl^c stretches (ν_1 and ν_2), a Mo–Cl$^{a(or\,b)}$ stretch (ν_3), a Cl^c–Mo–Cl$^{a(or\,b)}$ bending mode (δ_4) and a Mo–Mo stretch (ν_5). Their assignments are: ν_1 and ν_2 in the region 310–350 cm^{-1}; $\nu_3 = 246$ cm^{-1}; $\delta_4 = 110$ cm^{-1} and $\nu_5 = 220$ cm^{-1}. In analogous cluster compounds where bromide or iodide replace chloride, ν_5 is almost constant and so the strength of the metal–metal bond does not appear to be significantly affected by the identity of the halogen.

Four of the T_{1u} normal modes have also been observed in the infrared spectrum by Hartley and Wise.[7] The frequencies found were 221, 247, 330 and 351 cm^{-1}, in good agreement with those observed by Cotton and co-workers. Hartley and Wise pointed out that extensive mixing is likely to occur between normal vibrational modes and so a specific assignment of these frequencies may be of rather doubtful value.

Ten of the vibrational modes $(3A_{1g} + 3E_g + 4T_{2g})$ are expected to be active in the Raman spectrum of $[Mo_6Cl_8]Cl_6{}^{2-}$. Ten lines have been observed experimentally[7] and assigned as follows: 236, 318 and 402 cm^{-1} (A_{1g}); 179, 215 and 310 cm^{-1} (E_g); and 92, 195, 225 and 247 cm^{-1} (T_{2g}).

Comparative Properties of Molybdenum Chlorides

Chemical inter-relationships between the various chlorides of molybdenum are summarized in Table I and their structures and magnetic properties are given in Table II. $MoCl_5$ shows the paramagnetic behaviour expected for the presence of one unpaired electron. The magnetic moment of β-$MoCl_4$ is rather less than the value expected for two

TABLE I

Preparation and Inter-relationship of Molybdenum Chlorides

$$MoO_3 + SOCl_2 \xrightarrow[\text{several hr.}]{\text{reflux}} MoCl_6$$

$$2Mo + 5Cl_2 \xrightarrow{400-500°C} 2MoCl_5$$

$$2MoCl_5 + C_6H_6 \longrightarrow 2\alpha\text{-}MoCl_4 + C_6H_5Cl + HCl$$

$$2MoCl_5 + C_2Cl_4 \longrightarrow 2\alpha\text{-}MoCl_4 + C_2Cl_6$$

$$MoO_2 + CCl_4 \xrightarrow{250°C} \beta\text{-}MoCl_4 + CO_2$$

$$MoCl_3(s) + MoCl_5(g) \xrightarrow{250°C} 2\beta\text{-}MoCl_4(g)$$

$$4MoCl_5 + Mo \xrightarrow{235-255°C} 5\beta\text{-}MoCl_4$$

$$MoCl_5 \xrightarrow{H_2} \alpha\text{-}MoCl_3 + 2HCl$$

$$MoCl_5 \xrightarrow{\text{heat}} \text{'chloride-rich'} MoCl_3 \text{ (comp. } MoCl_{3.08}) \xrightarrow[\text{DMF}]{\text{extract with}} \beta\text{-}MoCl_3$$

$$Mo + MoCl_3 \xrightarrow{550-750°C} Mo_6Cl_{12}$$

$$2MoCl_3(s) \xrightarrow{\text{heat}} Mo_6Cl_{12}(s) + MoCl_4(g)$$

unpaired electrons and the α-form shows very much weaker paramagnetism. The long Mo,Mo distance in β-$MoCl_4$ precludes any significant bonding directly between these atoms in this modification. On the other hand, α-$MoCl_4$ is known to be isomorphous with $NbCl_4$, a diamagnetic compound which contains direct Nb,Nb bonds, and this suggests its abnormally low magnetic moment is due to Mo,Mo interaction. In α-$MoCl_3$, the Mo,Mo distance is quite short and its magnetic moment indicates extensive metal–metal bonding. A similar situation appears to exist in β-$MoCl_3$, although no structural information on this is available as yet. The shortest Mo,Mo distance and hence the strongest metal–metal bonding occurs in Mo_6Cl_{12}. This accounts for its diamagnetism.

These data on the chlorides of molybdenum illustrate the diagnostic value of magnetic susceptibility measurements in determining the extent of direct metal–metal bonding. It appears that such bonding is associated with a formal low oxidation state of the metal.

TABLE II

Structures and Magnetic Properties of Molybdenum Chlorides

Compound	Structure of Solid	Mo–Mo dist. (nm)	Magnetic moment (B.M.)*	
			Observed	Spin-only
$MoCl_6$	$MoCl_6$ molecules in which Mo is octahedrally coordinated	—	—	0
$MoCl_5$	Mo_2Cl_{10} molecules. Oct. coordination of Mo completed by bridging Cl	0·384	1·54[5]	1·73
β-$MoCl_4$	Derived from α-$TiCl_3$ with $\frac{3}{4}$ of metal positions occupied by Mo Octahedral coordination of Mo	0·350	2·31[5] 2·54[5] 2·42[10] 2·36[11]	2·83
α-$MoCl_4$	Isomorphous with $NbCl_4$ (this contains Nb–Nb bonds)	—	0·67 to 0·71[11]	2·83
α-$MoCl_3$	$AlCl_3$ layer lattice. Cubic close-packed Cl atoms. Some adjacent octahedral holes occupied by Mo–Mo pairs	0·276	0·49[5]	3·87
β-$MoCl_3$	$AlCl_3$ layer lattice. Hexagonal close-packed Cl atoms	—	0·96[5]	3·87
Mo_6Cl_{12}	$Mo_6Cl_8^{4+}$ clusters. Mo–Mo bonding	0·261	Diamagnetic	4·90

* The Bohr Magneton (B.M.) is the conventional unit of magnetic moment. Its equivalent in S.I. units is $9 \cdot 272 \times 10^{-24}$ Am^2, but in view of the convenient magnitude of the Bohr Magneton, this is used in the table and on subsequent occasions in the text.

Then the effective nuclear charge on the metal is low, its d-orbitals are expanded and can participate in bonding with other metal atoms.

Bonding

Metal–metal bonding could be described in terms of either the valence-bond or the molecular orbital theory. Cotton[8] has pointed out that the valence-bond approach is not helpful because the bonds which must be postulated are so bent or disproportionately concentrated in one spatial region that the simple concept of a two-centre bond is lost.

In an approximate M.O. treatment, some metal orbitals are involved in metal/ligand σ-bonding (π-bonding being neglected). The remaining orbitals are then combined according to group-theoretical procedures to form M.O.s which bind the metal atoms in the cluster.

In $[Mo_6Cl_8]^{4+}$, each molybdenum may be regarded as in a $+2$ oxidation state and can contribute four electrons to metal–metal bonds. There are therefore 24 bonding electrons associated with the 12 shortest metal–metal distances and the average bond order is 1·0. This can be compared with estimated bond orders in other species containing metal–metal bonds, e.g. in Re_3Cl_9 (Re,Re = 0·248 nm), there are 12 bonding electrons for the 3 metal–metal bonds and the bond order is 2·0; also, in $Re_2Cl_8^{2-}$(Re,Re = 0·224 nm), there are 8 electrons available for bonding the two rheniums which can therefore be considered as joined by a quadruple bond.

The relationship between bond length and magnetic properties in compounds containing metal–metal bonds has been discussed by Schäfer and Schnering.[9] The value of d(M–M), the metal–metal bond length, varies over a wide range, from 0·241 nm in $K_3[W_2Cl_9]$ to 0·331 nm in α-NbI_4, in diamagnetic compounds. In contrast, $K_3Cr_2Cl_9$, in which d(Cr–Cr) = 0·312 nm, is paramagnetic. It is not therefore a simple case of diamagnetism being associated with a M–M bond of particular shortness. The absolute value of d(M–M) appears to depend strongly on the dimensions of the crystal lattice. For example, in β-$MoCl_4$, α-$MoCl_3$ and Mo_6Cl_{12}, the metal atoms occupy octahedral holes in the chloride ion lattice. d(Mo–Mo) varies with the distance, $\bar{d}$, between neighbouring octahedral holes,

TABLE III

Relative Shortening of Mo–Mo Distance in Molybdenum
Chlorides

	d(Mo–Mo) (nm)	$\bar{d}$ (nm)	$\Delta d/\bar{d}$
β-$MoCl_4$	0·350	0·350	0·0
α-$MoCl_3$	0·276	0·323	0·145
Mo_6Cl_{12}	0·261	0·306	0·147

as shown in Table III. The ratio $\Delta d/\bar{d}$ where $\Delta d = \bar{d} - d$(M–M) expresses the relative shortening of the Mo–Mo bonds. The greater the value of this ratio, the more extensive the metal–metal bonding. Similar calculations show that the corresponding ratio for α-NbI_4 is 0·138 and for $K_3Cr_2Cl_9$ is zero. This ratio thus is a more meaningful quantity to relate to magnetic properties than the actual magnitude of d(M–M).

References

1. COTTON, F. A., *Quart. Rev.* 389 (1966).
2. SHELDON, J. C., *J. Chem. Soc.* 1007 (1960).
3. SHELDON, J. C., *J. Chem. Soc.* 4183 (1963).
4. COTTON, F. A., and CURTIS, N. F., *Inorg. Chem.* **4**, 241 (1965).
5. SCHÄFER, H., SCHNERING, H. G., TILLACK, J., KUHNEN, F., WÖHRLE, H., and BAUMANN, H., *Zeit. anorg. allgem. Chem.* **353**, 281 (1967).
6. COTTON, F. A., WING, R. M., ZIMMERMAN, R. A., *Inorg. Chem.* **6**, 11 (1967).
7. HARTLEY, D., and WISE, M. J., *Chem. Comm.* 912 (1967).
8. COTTON, F. A., *Inorg. Chem.* **4**, 334 (1965).
9. SCHÄFER, H., and SCHNERING, H. G., *Angew. Chem.* **76**, 833 (1964).
10. KEPERT, D. L., and MANDYCZEWSKY, R., *Inorg. Chem.* **7**, 2091 (1968).
11. BROWN, T. M., and McCANN, E. L., *Inorg. Chem.* **7**, 1227 (1968).

CHAPTER 18

POTASSIUM RHENIUM HYDRIDE

Introduction

Hydrogen forms compounds with almost all the other known elements. In the case of the highly electropositive metals, which form salt-like hydrides, and in the case of non-metals, which form covalent molecules, the behaviour of hydrogen is well understood. Much less is known about the bonding in compounds containing transition metals and hydrogen and considerable interest has been shown in the relatively few compounds which appear to contain directed covalent bonds between hydrogen and the metal.

The first binary compound of hydrogen with a transition metal to be synthesized was copper hydride, CuH, described by Wurtz in 1844.[1] This is produced as a red-brown precipitate by the reduction of aqueous copper sulphate with hypophosphorous acid. No analogous reaction takes place with other transition metals. Binary hydrides of these are generally obtained by direct combination between hydrogen and the metal. The compounds formed are typically semi-metallic in their properties and frequently non-stoichiometric in their composition.

In the early 1930s, the carbonyl hydrides, $H_2Fe(CO)_4$ and $HCo(CO)_4$ were discovered. These are now known to contain direct metal–hydrogen bonds. Other complex hydrides containing similar bonds have since been synthesized. These include the cyclopentadienyl carbonyl hydrides like $(C_5H_5)Cr(CO)_3H$ and the tertiary phosphine hydrides like *trans*-$[PtHCl(PEt_3)_2]$. All these compounds have characteristic proton magnetic resonance spectra showing large chemical shifts to high fields. This behaviour typifies hydrogens which are directly linked to metal atoms.

There are only two hydrides known in which hydrogen is the sole ligand bound to the transition metal. These are dipotassium enneahydridorhenate, K_2ReH_9, and its technetium analogue, K_2TcH_9.

In the rhenium compound it has proved difficult to establish the number of hydrogens bound to one metal atom. From the viewpoint of the analytical chemist, determination of the hydrogen content is not easy because hydrogen is such a small percentage of the total composition. Again, some techniques for structure determination such as X-ray crystallography are not particularly helpful because they cannot show the positions of the hydrogen atoms directly. It is not surprising, therefore, that the composition of potassium rhenium hydride remained in doubt for some years and it was not until the structure was elucidated by neutron diffraction that the true stoichiometry was established.

165

Preparation

Dipotassium enneahydridorhenate is prepared by the reduction of potassium perrhenate, $KReO_4$, with potassium metal in aqueous ethylenediamine solution.[2] The product is contaminated with impurities such as ReO_2 and Re, unchanged $KReO_4$ and KOH. The ReO_2, Re and $KReO_4$ remain undissolved when the impure product is extracted with KOH and can be removed by filtration. The complex hydride is purified by repeated solution in methanol and precipitation by the addition of 2-propanol. It is finally obtained as a white, deliquescent solid which undergoes slow decomposition on exposure to the atmosphere.

In the final stages of purification, it is convenient to run the infrared spectrum as a measure of the impurities present. The product shows bands at 1846 and 735 cm^{-1} which are believed to be due to rhenium–hydrogen bonds. Potassium perrhenate shows a triplet at 915, 920 and 935 cm^{-1} and the strongest band of these, at 915, can be used as a sensitive test for this impurity.

Stoichiometry

The product of the reduction of $KReO_4$ with potassium was first tentatively identified as a hydrated potassium rhenide, $KRe \cdot 4H_2O$, containing the rhenide ion, Re^-, in which rhenium shows an oxidation state of -1. It was soon recognized that its properties were not consistent with this formulation. The inference from the infrared spectrum that Re–H bonds are present has been confirmed by N.M.R. spectroscopy.

The proton resonance spectrum, measured at 60 MHz on a solution of the solid in 50% aqueous potassium hydroxide, shows two peaks, separated by 14·7 ppm. The low-field peak is very large and is due to the hydroxo group protons. The high-field peak, which shows a shift of $+13·3$ ppm with respect to the proton resonance of pure water, is due to hydrogens bound directly to rhenium (hydridic protons). In measurements on two different samples, the hydrogen:rhenium ratio was found to be 8·12 and 7·96. This evidence was, for some time, taken to establish that the compound should be formulated as a complex hydride of formula, K_2ReH_8, with the rhenium in a $+6$ oxidation state.

Elemental analyses of the compound and quantitative studies of its behaviour on oxidation were in broad agreement with this conclusion. The rhenium content of the product has been determined by oxidation with hydrogen peroxide to potassium hydroxide and perrhenate. The potassium hydroxide can be quantitatively analysed by potentiometry with standard acid and subsequently the rhenium content of the neutralized solution determined gravimetrically by the precipitation of tetraphenylarsonium perrhenate. In this way, the atomic ratio, K:Re, has been shown to be close to 2 for several different samples.

Oxidation of an alkaline solution of the solid with hypochlorite produces 2·5 molecules of hydrogen for each atom of rhenium. In acid solution, the solid decomposes with the evolution of hydrogen and the precipitation of rhenium metal, 5 molecules of hydrogen being evolved for each atom of rhenium. On the basis of the formulation of the compound as K_2ReH_8, these reactions were interpreted according to the equations:

$$ReH_8{}^{2-} + 2H^+ \rightarrow Re^0 + 5H_2$$

$$\text{and} \quad K_2ReH_8 + 6NaClO \rightarrow KReO_4 + 2·5H_2 + KOH + H_2O + 6NaCl$$

Experimental values for the molecular ratio of hypochlorite to rhenium hydride varied between 5·9 and 6·15 as compared with the theoretical requirement of 6.

The validity of these conclusions must depend on the purity of the sample analysed and the nature of any impurities present.

There is one property of potassium rhenium hydride which is not consistent with its formulation as K_2ReH_8. Thus the N.M.R. spectrum of an alkaline solution of the hydride and magnetic susceptibility measurements on the solid show that the compound is diamagnetic.[3] This property cannot be reconciled with the presence of an anion containing Re(VI) for the metal should have one unpaired electron in this oxidation state and its complexes should be paramagnetic. A possible explanation of the diamagnetism in terms of a dimeric ion formed by a metal–metal bond between the two rhenium atoms (analogous to that found in polynuclear carbonyls such as $Mn_2(CO)_{10}$) is very unlikely in view of the X-ray crystallographic data.[4] The smallest Re,Re distance is 0·551 nm, a value much greater than that expected should metal–metal bonding exist.

Structure

X-ray analysis shows that each rhenium atom is surrounded by nine potassium ions, six of which are arranged in a trigonal prism with the remaining three beyond the centres of the rectangular faces. From the Re,K distances it appears that there are two non-equivalent groups of rhenium atoms, designated Re_1 and Re_2. These appear to differ in a crystallographic sense rather than in a chemical sense. For Re_1, there are six potassiums at a distance of 0·3701 ± 0·0008 nm, and three at 0·3939 ± 0·0015 nm. For Re_2, there are three potassiums at 0·3623 ± 0·0013 nm, and 6 at 0·4003 ± 0·0007 nm.

The study of the hydride by neutron diffraction[5] has established unequivocally that there are actually nine hydrogen atoms combined with each rhenium. Six of these are located at the corners of a trigonal prism, with the rhenium atom at the centre, and the remaining three are beyond the centres of the rectangular faces of the prism (Fig. 18.1). The stoichiometry of the compound must be K_2ReH_9. Its diamagnetism can now be understood because Re(VII) must be present and this contains no unpaired electron.

The conclusion that nine and not eight hydrogens are bound to each rhenium prompted a re-investigation of the composition of the hydride by chemical analysis. The hydrogen

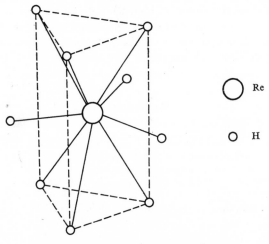

Re

H

FIG. 18.1. The shape of the ReH_9^{2-} ion.

content may be directly determined by the thermal decomposition of a known weight. *In vacuo*, hydrogen gas and potassium and rhenium metals are formed. The amount of hydrogen evolved may be measured volumetrically. Mass spectrometric analysis of gas produced in this way has shown that it contains a small proportion of methane. This must originate from some organic impurity in the hydride, presumably ethylenediamine. Duplicate values reported for the atomic ratio H:Re are 8·70 and 8·67:1. In view of the presence of some impurity and the difficulty of making highly accurate analyses for hydrogen because of the small contribution which this element makes to the formula weight, the analytical data are in reasonable agreement with the ratio of 9:1 required for K_2ReH_9.

Evidence for Other Hydrides of Rhenium

In 1937, Lundell and Knowles[6] showed that a rhenium species with reducing properties was obtained by reduction of an acid perrhenate solution with zinc amalgam. The formal oxidation state of this was determined by addition of the reduced solution to excess ferric alum solution, followed by titration with standard dichromate of the ferrous iron produced by oxidation of rhenium back to perrhenate. Eight equivalents of ferric iron were required to oxidize each rhenium and this was believed to show that the reduced species was the rhenide ion, Re^-.

Later investigations by Ginsberg and Koubek[7] have shown that this is unlikely to be so. The ultraviolet spectrum of a simple rhenide ion should follow the pattern of the spectra of halide ions. These are characterized by intense charge-transfer bands with maxima at 181, 200 and 226 nm for Cl^-, Br^- and I^- respectively. Rhenium, with an atomic weight greater than that of iodine, should show charge-transfer absorbance at wavelengths greater than 226 nm. In fact, the absorbance of 'rhenide' solutions in sulphuric acid is very low at any wavelength above 210 nm.

It has been shown by polarography that eight electrons per rhenium atom are required to oxidize 'rhenide' back to perrhenate. If the reduced species is not Re^-, then it is most likely to be a hydrido-complex of an intermediate oxidation state of rhenium. It cannot be ReH_9^{2-} because oxidation of this to perrhenate would be a 9-electron reaction. A definitive confirmation of hydridic nature could come from an observation of the proton resonance line characteristically found at high field. So far, it has not proved possible to obtain a solution in which the concentration of reduced species is above 10^{-3} mol dm^{-3} and this is too dilute to allow observation of the expected line. The nature of the reduced species and the oxidation state of rhenium in it remain obscure at the present time.

References

1. WURTZ, A., *Compt. Rend.* **18**, 702 (1844).
2. GINSBERG, A. P., MILLER, J. M., and KOUBEK, E., *J. Amer. Chem. Soc.* **83**, 4909 (1961).
3. KNOX, K., and GINSBERG, A. P., *Inorg. Chem.* **1**, 945 (1962).
4. KNOX, K., and GINSBERG, A. P., *Inorg. Chem.* **3**, 555 (1964).
5. ABRAHAMS, S. C., GINSBERG, A. P., and KNOX, K., *Inorg. Chem.* **3**, 558 (1964).
6. LUNDELL, G. E. F., and KNOWLES, H. B., *J. Res. Nat. Bur. Stan.* **18**, 629 (1937).
7. GINSBERG, A. P., and KOUBEK, E., *Zeit. anorg. all. Chem.* **315**, 278 (1962).

CHAPTER 19

IRON CARBONYLS

Introduction

The properties of carbon monoxide as a ligand are well established from the synthesis and study of its binary complexes with many transition metals. In addition, numerous mixed complexes are known containing CO and a second ligand such as nitric oxides substituted phosphines, arsines and stibines. These ligands all have the property of stabilizing low or zero oxidation states of the transition metal with which they are combined.

The first-known binary carbonyl, nickel tetracarbonyl, $Ni(CO)_4$, was synthesized in 1890, and, in the following year, iron pentacarbonyl, $Fe(CO)_5$, was made independently by Mond[1] and Berthelot.[2] Many other carbonyls, including several polynuclear complexes, have subsequently been prepared. The Mond process for the purification of nickel is the prime example of the industrial exploitation of the properties of carbonyls. Nickel tetracarbonyl and other metal carbonyls decompose into their components on heating and it is possible to prepare metals in a state of high purity in this way.

Carbonyls and related compounds are intermediates in various processes where transition metals act as catalysts. For example, the Fischer–Tropsch synthesis of hydrocarbons and alcohols from carbon monoxide and hydrogen involves the use of cobalt as a catalyst and the reaction is believed to proceed via the formation of a cobalt carbonyl hydride. The primary stage in the synthesis is the reaction between carbon monoxide adsorbed on the surface of the metal and hydrogen. Another well-known example is the catalytic activity displayed by iron pentacarbonyl in the hydrogenation and isomerization of vegetable oils.

The formation and properties of metal carbonyls have stimulated great interest because of the novelty of bonding between carbon monoxide molecules and metal atoms in a low or zero oxidation state. This is unexpected in view of the lack of basic properties of carbon monoxide and it is difficult, at first sight, to explain why a metal atom should show any marked tendency to accept electrons for in so doing it would accumulate negative charge in contravention of the electroneutrality principle.

Iron forms three binary carbonyls: iron pentacarbonyl, $Fe(CO)_5$; di-iron enneacarbonyl, $Fe_2(CO)_9$; and tri-iron dodecacarbonyl, $Fe_3(CO)_{12}$. The two polynuclear carbonyls show additional features of interest, beyond those already mentioned which apply to carbonyls in general, because their structures involve direct metal–metal bonding and the functioning of the carbon monoxide molecules as bridging groups as well as monodentate ligands.

169

IRON PENTACARBONYL, $Fe(CO)_5$

Preparation

The direct reaction between finely-divided iron and CO gas under pressure and at an elevated temperature produces iron pentacarbonyl. This is a straw-coloured liquid, b.p. 103°C. The liquid freezes at −20°C.

Reactions

On irradiation with ultraviolet light, $Fe(CO)_5$ is converted to $Fe_2(CO)_9$. Reaction with aqueous sodium hydroxide produces the sodium salt, $Na[HFe(CO)_4]$, which yields the parent carbonyl hydride, $H_2Fe(CO)_4$, on acidification.

In reactions with a number of other ligands, the carbon monoxide is partly replaced by them. For example, reaction with cyclopentadiene gives the mixed cyclopentadienyl iron carbonyl, $[C_5H_5Fe(CO)_2]_2$. Triphenylphosphine reacts in a similar way to give $(C_6H_5)_3 \cdot PFe(CO)_4$ and $[(C_6H_5)_3P]_2Fe(CO)_3$, and bicyclohepta-2,5-diene gives the mixed complex, $C_7H_8Fe(CO)_3$.

When a mixture of $Fe(CO)_5$ and triethylamine is heated to 80°C under a nitrogen atmosphere for 10 hr., the compound $[(C_2H_5)_3NH][HFe_3(CO)_{11}]$ is formed. On acidification of this with 1:1 hydrochloric acid/water mixture, $Fe_3(CO)_{12}$ is formed.

Structure

From an electron diffraction study, Ewens and Lister[3] concluded that the $Fe(CO)_5$ molecule has a trigonal bipyramidal shape. The carbon atoms are bound to the iron atom and the bond lengths are Fe,C = 0·184 ± 0·003 nm and C,O = 0·115 ± 0·004 nm. All possible alternative models, such as a square-based pyramid, a regular plane pentagon, or a trigonal bipyramid with oxygens linked to the iron, gave a poorer correlation between theoretical predictions and experimental results.

A second electron diffraction investigation[4] has led to the conclusion that the Fe,C bonds may be differentiated according to their lengths, the axial being shorter (0·1797 nm) than the equatorial (0·1842 nm). This difference appears to be significant[5] but unexpected in view of X-ray diffraction data on the solid[6,7] which do not provide any reason to suppose the axial bonds are shorter but, if anything, suggest they may be slightly longer than the equatorial. Parameters of the molecule in the gas and solid phases need not be identical and there could be some structural change accompanying the phase change which accounts for the different data.

The structure deduced by Donohue and Caron[7] from X-ray diffraction is illustrated in Fig. 19.1. The difference in the Fe,C bond lengths is regarded by them as not significant.[8]

The different investigators agree, however, that the molecular shape is close to that of a regular trigonal bipyramid. This throws some doubt on previous measurements of the dipole moment of $Fe(CO)_5$ which had been variously reported as 0·62D[9] and 0·81D,[10] both of these values being far too high for such a symmetrical molecule. Horrocks and DiCarlo[11] have measured the dielectric constant and dielectric loss of pure liquid $Fe(CO)_5$ at microwave frequencies and found values of 0·15 and 0·10D for the dipole moment at wavelengths 1·25 and 3·22 cm respectively. These are quite consistent with the known

170

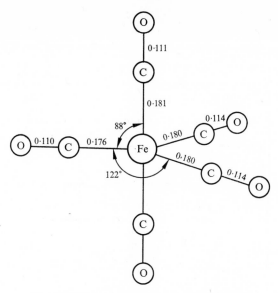

FIG. 19.1. The structure of iron pentacarbonyl.

structure of $Fe(CO)_5$ and it is probable that the earlier dipole moment values are inaccurate because of an underlying assumption that atom polarization could be taken as zero.

Vibrational Spectra

The first reported studies of the infrared and Raman spectra of $Fe(CO)_5$ were incomplete[13,12] and probably marred by the presence of bands due to impurities. The spectra were interpreted using a model of either C_{4v} (square pyramid[12] or D_{3h} (trigonal bipyramid)[13] symmetry. Although this is not now believed to be correct, the model of C_{4v} symmetry was at one time quite consistent with the dipole moment data then available.

For some years, a decision between these two possibilities could not be reached due to the experimental difficulty in obtaining a Raman spectrum. This prevented to complete assignment of the fundamental vibrations of the molecule. The first attempts to induce a Raman spectrum of $Fe(CO)_5$ were made by irradiation of the compound with a mercury vapour source using the mercury lines at 435·8 and 546·1 nm for excitation. The carbonyl absorbs light at these wavelengths and is converted to $Fe_2(CO)_9$. Photodecomposition was reduced by changing to irradiation with the sodium vapour doublet at 589·0 and 589·6 nm. Even so, the only Raman lines which could be indentified as C–O stretching frequencies were those observed at 1995 cm^{-1} on samples of the pure carbonyl. Other, lower frequencies were observed on solutions of the carbonyl in an inert solvent, cyclohexane. Eventually, a complete Raman spectrum was obtained[15] by excitation with helium vapour radiations at 667·8 and 706·52 nm in conjunction with liquid filters containing carbocyanine dyes to absorb shorter wavelength radiation and so prevent decomposition.

The selection rules for D_{3h} and C_{4v} symmetries and the observed frequencies of the Raman shifts are given in Table I. The results clearly support a molecular model of D_{3h} symmetry.

171

TABLE I

Raman Spectrum of Iron Pentacarbonyl in the Liquid State

Approx. description of vibrational modes	C–O stretch	Fe–C–O bend	Fe–C stretch	C–Fe–C bend
Raman shift (cm^{-1})	1984 2031 2114	492 558 640 753	377 414 441	68 104 112
Number of observed shifts	3	4	3	3
Number expected for D_{3h} symmetry	3 $(2A_1' + E')$	4 $(2E'' + 2E')$	3 $(2A_1' + E')$	3 $(2E' + E'')$
Number expected for C_{4v} symmetry	4 $(2A_1 + B_1 + E)$	6 $(B_1 + B_2 + A_1 + 3E)$	4 $(B_1 + 2A_1 + E)$	5 $(A_1 + B_1 + B_2 + E)$

The definitive infrared spectroscopic study is that of Edgell, Wilson and Summitt[16] who examined the high resolution spectra of the liquid and vapour over the range 4200–250 cm^{-1} and also the solid as a film at liquid nitrogen temperatures within the region 700–300 cm^{-1}. High purity of the sample was ensured by repeated recrystallizations and confirmed by gas chromatography.

Four main types of vibrational modes can be recognized. Firstly, there are four C–O stretching modes, expected to occur at high frequencies. They are represented in Fig. 19.2. The infrared band at 2034 cm^{-1} is coincident with the Raman line at 2031 cm^{-1} and has been identified with the E' mode (ν_{10}). This is also observed in the infrared spectrum as an overtone band at 4054 cm^{-1} ($2\nu_{10}$). The other band, at 2014 cm^{-1}, must originate with the A_2'' mode (ν_6). Raman lines at 2114 and 1984 cm^{-1} arise from ν_1 and ν_2 and comparisons with the spectra of other metal carbonyls have led to the identification $\nu_1 = 2114$, $\nu_2 = 1984$.

TABLE II

Infrared Spectrum of Iron Pentacarbonyl in the Vapour State

Approx. description of vibrational modes	C–O stretch	Fe–C–O bend	Fe–C stretch	C–Fe–C bend
Frequency (cm^{-1})	2014 2034	544 620 646	431 474	104
Number of observed fundamentals	2	3	2	1
Number expected for D_{3h} symmetry	2 $(A_2'' + E')$	3 $(A_2'' + 2E')$	2 $(A_2'' + E')$	3 $(A_2'' + 2E')$
Number expected for C_{4v} symmetry	3 $(2A_1 + E)$	3 $(2A_1 + E)$	4 $(A_1 + 3E)$	3 $(A_1 + 2E)$

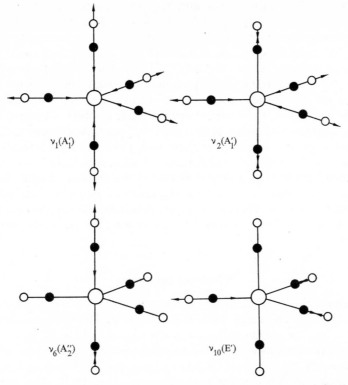

FIG. 19.2. The four C–O stretching modes in iron pentacarbonyl.

The other three kinds of vibrational modes are: Fe–CO stretching, in which the CO groups tend to move as rigid units; C–O bending, with some deformation of the Fe–C–O bond angles; and C–Fe–C bending modes, associated with slight deformation of the Fe–C–O angles. The frequencies observed in the infrared spectrum of $Fe(CO)_5$ vapour are given in Table II and, like the bands in the Raman spectrum, they are in accordance with D_{3h}, and not C_{4v}, symmetry.

The infrared spectrum of iron pentacarbonyl provides valuable insight into the nature of the iron–carbon and carbon–oxygen bonds. The C,O stretching frequencies are rather lower than the value (2146 cm^{-1}) in carbon monoxide itself. This is consistent with a decrease in the order of the C,O bond in the carbonyl. Significant decrease in the C,O stretching frequency is a general phenomenon in metal carbonyls and it typically occurs within the range 2050–1900 cm^{-1}. In polynuclear carbonyls, additional absorption occurs between 1800 and 1900 cm^{-1}. This is identified with bridging carbonyl groups. The bridging action of a carbonyl group in joining two metal atoms reduces the bond order even further below that in terminal carbonyl groups.

N.M.R. Spectra

Unequivocal support for the trigonal bipyramidal shape of the $Fe(CO)_5$ molecule should come from the ^{13}C N.M.R. spectrum. Thus a trigonal bipyramid should show fine structure of the A_3B_2X type, that is, two lines with an intensity ratio of 3:2. Alternatively, the spectrum for a square-based pyramidal molecule should be of the A_4BX type and show two lines with an intensity ratio of 4:1.

In fact, the ^{13}C N.M.R. spectrum is reported[17] to show only a single resonance at ordinary temperature and below. Similarly, a single line spectrum is obtained when measuring the ^{17}O nuclear resonance. The failure to observe more than one line could arise if the chemical shift between the axial and equatorial carbon monoxide groups were smaller than the resolution of the spectrometer used. It has been estimated that if this is the reason, the chemical shift must be less than 0·25 ppm. Alternatively, all CO groups could be equivalent if rapid exchange of these took place. An intermolecular mechanism seems to be ruled out in view of exchange studies[19] using radio-carbon which have shown that such exchange proceeds at a very slow rate in the absence of light. Intramolecular exchange could occur either by a C_{4v} to a D_{3h} inversion or *via* an octahedral model in which one position is vacant. There is no experimental data to support this theory and it is not possible to account, in an entirely satisfactory way, for the single line ^{13}C and ^{17}O N.M.R. spectra.

Mössbauer Spectrum

^{57}Fe has nuclear spin, $I = 1/2$, in the ground state and $I = 3/2$ in the first excited state. These are separated by an energy difference of 2 · 31 fJ. The natural abundance of this isotope is 2·17% and the energy difference is low enough for the Mössbauer effect to be observable in iron compounds even at ordinary and higher temperatures.

The Mössbauer spectrum of $Fe(CO)_5$ shows quadrupole splitting and consists[19] of a well-defined doublet at liquid nitrogen temperature (Fig. 19.3). The separation, ΔE, = 2·60 mm sec^{-1}. The doublet arises from the splitting of the excited state due to a field gradient at the nucleus and indicates some asymmetry in the *s*-electron distribution within the iron atom.

As well as quadrupole splitting, another parameter of major interest in iron compounds is the nuclear isomer shift. This is the displacement from zero velocity of the centroid of the Mössbauer spectrum and occurs because the electronic charge density at the nucleus interacts with the nuclear charge in the excited and ground states to different extents, depending on the chemical environment of the iron atom. The only electrons with a finite density at the nucleus are those in *s*-orbitals. The outermost *s*-electrons are involved in bonding and the charge density due to them is affected by the chemical environment. Mössbauer spectroscopic studies on various iron compounds have established that values of the isomer shift are diagnostic of the oxidation state of the metal.

The isomer shift actually measured is the difference in *s*-electron density at the nuclei of the iron atoms of the sample and of the source. For $Fe(CO)_5$, the isomer shift is very small[20] (+0·035 ± 0·043 mm sec^{-1} with respect to 310 stainless steel as reference). This has been interpreted in terms of essentially covalent bonding between the ligands and iron and of the dsp^3 orbital hybridization implied by the trigonal bipyramidal structure. The increase in 4*s*-electron density associated with this hybridization would be expected to change the isomer shift towards negative values. This effect appears to be just about counterbalanced by the concomitant increase in 3*d*-electron density which produces increased shielding of the nuclear charge with respect to the 4*s*-electrons, a consequent expansion of the 4*s*-orbitals and a reduction of *s*-electron density at the nucleus almost to the value found for metallic iron.

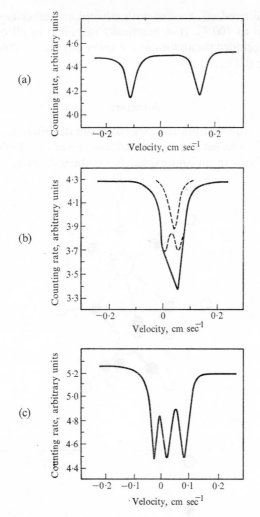

FIG. 19.3. Mössbauer spectra of iron carbonyls: (a) iron pentacarbonyl, (b) di-iron enneacarbonyl (this contains some iron(III) which produces absorption at 0·05 cm sec⁻¹ and the dotted lines indicate the resolution of the spectrum into its two components), and (c) tri-iron dodecacarbonyl (reproduced with permission from Herber, R. H., Kingston, W. R., and Wertheim, G. K., *Inorg. Chem.* **2**, 153 (1963)).

DI-IRON ENNEACARBONYL, $Fe_2(CO)_9$

Preparation

This was originally prepared[21] by exposing an acetic acid solution of $Fe(CO)_5$ to sunlight. The maximum yield reported in this reaction was 30%. A modified procedure[22] which is recommended for much higher yield (74–91%) involves the use of artificial light and water cooling of the reaction mixture. An acetic acid solution of $Fe(CO)_5$ is irradiated for 24 hr by a mercury vapour lamp. During this time, $Fe_2(CO)_9$ separates out as orange

175

crystals which can be filtered off, washed with ethanol and ether and dried *in vacuo*. The compound is stable up to 100°C. It is practically insoluble in all organic solvents. This lack of solubility has precluded the application of many physical methods in the investigation of the properties of $Fe_2(CO)_9$.

Structure

The positions of the atoms within the molecule were definitely established in the X-ray crystallographic study[23] carried out by Powell and Ewens in 1939. Each iron atom is surrounded by six carbons in an approximately octahedral environment (Fig. 19.4). The

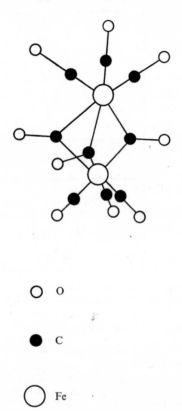

○ O

● C

◯ Fe

FIG. 19.4. The structure of di-iron enneacarbonyl.

carbonyl groups present are differentiated according to their functioning as terminal (bound to one iron atom) or as bridging groups (linking two iron atoms). The C,O inter-nuclear distance in a terminal group is 0.115 ± 0.05 nm; that in a bridging group is 0.130 ± 0.005 nm. A notable feature of the structure is the short Fe,Fe distance of 0.246 nm. This, and the observed diamagnetism of the compound are accounted for by postulating the formation of a metal–metal bond involving the two unpaired electrons, originally located one on each of the two iron atoms.

The Mössbauer spectrum of $Fe_2(CO)_9$ (Fig. 19.3) shows[20] the expected quadrupole splitting but this, 0.541 ± 0.043 mm sec^{-1}, is only 20% of that observed in $Fe(CO)_5$.

This means the total field gradient at the nucleus must be small and this is consistent with the structure, as determined by X-ray analysis, which shows only a small departure from octahedral symmetry. The major contribution to the field gradient is probably from the two different types of carbonyl ligands with the iron–iron bond having only a small influence. The observed isomer shift of 0.282 mm sec^{-1} corresponds with a decrease in electron density due to $4s$-electrons at the nucleus compared with $Fe(CO)_5$. The two peaks in the spectrum do not have equal intensities. This asymmetry is not the result of any difference in the environment of the two iron atoms but appears to be due entirely to preferential orientation of the molecules in the crystal.[24]

Tri-iron Dodecacarbonyl, $Fe_3(CO)_{12}$

Preparation

This compound was first synthesized[25] in 1906 by the thermal decomposition of $Fe_2(CO)_9$. It has also be made by two other methods, the oxidation of the $HFe(CO)_4^-$ ion[26] and the reaction between $Fe(CO)_5$ and triethylamine followed by acidification of the product.[27]

A solution containing $HFe(CO)_4^-$ ions is prepared by the treatment of $Fe(CO)_5$ in methanol with aqueous sodium hydroxide under a nitrogen atmosphere:

$$Fe(CO)_5 + 3OH^- \rightarrow HFe(CO)_4^- + CO_3^{2-} + H_2O$$

This solution is then oxidized with manganese dioxide:

$$3HFe(CO)_4^- + 3MnO_2 + 3H_2O \rightarrow Fe_3(CO)_{12} + 3Mn^{2+} + 9OH^-$$

Oxidation proceeds for a period of 1–2 hr, the solution becoming dark red in colour. Excess MnO_2 is destroyed by adding ferrous sulphate and $Fe_3(CO)_{12}$ is precipitated as a dark green crystalline solid on the addition of sulphuric acid (1:1).

Triethylamine and $Fe(CO)_5$ react slowly when heated together at 80°C in a nitrogen atmosphere with the formation of $[(C_2H_5)NH][HFe_3(CO)_{11}]$ which separates as a red solid.

$$3Fe(CO)_5 + (C_2H_5)_3N + 2H_2O \rightarrow [(C_2H_5)_3NH][HFe_3(CO)_{11}] + 2CO_2 + 2CO + H_2.$$

When this product is treated with HCl (1:1), $Fe_3(CO)_{12}$ is formed.

Tri-iron dodecacarbonyl can be purified by sublimation at 60°C under low pressure or by Soxhlet extraction with n-pentane under nitrogen. It is slowly oxidized to Fe_2O_3 by air and must be preserved under nitrogen. Heating above 140°C causes decomposition to metallic iron and CO.

Structure

In 1930, Hieber and Becker[27] showed from measurements of the freezing-point depression of solutions of iron 'tetracarbonyl' in iron pentacarbonyl as solvent, that the 'tetracarbonyl' is indeed trimeric. Following this discovery, at least six different structures have, from time to time, been proposed for the $Fe_3(CO)_{12}$ molecule. Evidence from physical studies has been conflicting and it was not until very recently[28] that the controversy over the stereochemistry of this compound was settled.

Two of the models favoured in the past have a collinear arrangement of iron atoms which are joined together by carbonyl bridges. One of these (I) is of D_{3d} symmetry and contains three bridging carbonyl groups joining the central iron atom to each of two $Fe(CO)_3$ groups so that each iron is octahedrally bound by six CO groups. This may be referred to as the 3-3-3-3 model. The other (II) is a 4-2-2-4 model and contains two bridging carbonyl groups which link the central iron atom to each of the two $Fe(CO)_4$ groups. This model has D_{2d} symmetry and the four bridging carbonyls are tetrahedrally arranged round the central iron atom.

Support for either (I) or (II) came from infrared spectroscopic data on a saturated solution of $Fe_3(CO)_{12}$ in toluene.[29] There is evidence for terminal carbonyl (strong bands at 2020 and 2043 cm^{-1}) and bridging carbonyl (very weak bands at 1833 cm^{-1}) groups. Although weak absorption bands in the 1800 cm^{-1} region have been observed by other investigators, some doubt has been expressed on whether or not these are fundamental frequencies or overtone or combination bands. In the spectrum of the carbonyl, examined as a potassium bromide disc, two bands in this region are clearly resolved[30] and their intensities, relative to those of the bands just above 2000 cm^{-1}, are much greater in the solid state than in solution. This evidence certainly suggests that the very weak bands observed for $Fe_3(CO)_{12}$ in solution are not fundamental and hence that bridging groups are absent.

One of the first structural analyses[31] of the solid by X-ray diffraction, reported in 1957, was believed to favour a centrosymmetric model with collinear metal atoms. At about the same time, however, Dahl and Rundle[32] deduced from their X-ray studies a triangular arrangement of metal atoms, with Fe,Fe distances between 0·275 and 0·285 nm. Their experimental data could not be accounted for by a linear arrangement of iron atoms but they were unable, because of an order–disorder phenomenon in the crystal, to reach a complete description of the structure. However, since the polarized single crystal infrared spectrum[30] of $Fe_3(CO)_{12}$ shows a strong band at 1875 cm^{-1} (within the frequency range where absorption by bridging carbonyls is expected), it appeared probable that the structure in the crystalline state consists of a symmetrical triangle of iron atoms, with some of the carbonyls acting as terminal and some as bridging groups.

In 1963, the Mössbauer spectra of solid samples of $Fe_3(CO)_{12}$ were reported[20,33] for the first time. These show (Fig. 19.3) three resonance maxima of equal intensity. The presence of three iron atoms which all differ electronically is very unlikely and the outer two resonance lines, located at $+0·847$ and $-0·287$ mm sec^{-1} respectively, were interpreted as being due to a marked quadrupole splitting of the resonance of two equivalent iron atoms caused by a large electric field gradient at their nuclei. The central line, at $+0·208 \pm 0·043$ mm sec^{-1} then represents a single, unsplit resonance due to the third iron atom possessing a very small electric field gradient at its nucleus. This kind of spectrum is, in fact, that to be expected for a linear arrangement of iron atoms if the central atom is in an environment of high symmetry, i.e. essentially octahedral, and the outer atoms are in an environment of lower symmetry. It is not consistent with a trigonal arrangement in which the Fe,Fe distances are equal because all three metal atoms would then have an equivalent environment.

The contradictory interpretations of Mössbauer and X-ray data stimulated further efforts to solve the problem of the structure of $Fe_3(CO)_{12}$. It is now firmly established[28] that the molecule in the crystalline solid has the iron atoms situated at the corners of an isosceles triangle. This conclusion was reached by Dahl and Blount[34] by analogy with the structure of the closely-related anion, $HFe_3(CO)_{11}^-$. In this, the three iron atoms are in a triangular arrangement and it has been inferred, from stereochemical and other considerations, that the hydrogen is located so as to act as a symmetrical bridge between two metal atoms. Probably the $Fe_3(CO)_{12}$ molecule is derived from this by the replacement of hydrogen by a carbonyl group. The known structure[35] of $Fe_3(CO)_{11}P(C_6H_5)_3$ also supports this model for $Fe_3(CO)_{12}$.

Wei and Dahl[28] have now determined the structure of $Fe_3(CO)_{12}$ unambiguously by X-ray diffraction on single crystals. It is shown in Fig. 19.5 and may be regarded as com-

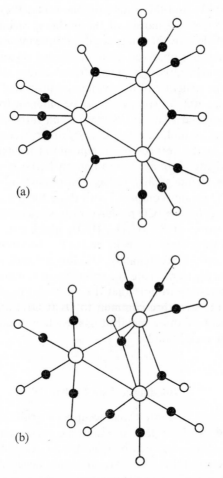

FIG. 19.5. Proposed structures of $Fe_3(CO)_{12}$: (a) in solution, and (b) in the solid state.

posed of an $Fe(CO)_4$ group symmetrically coordinated by only Fe,Fe bonds to an $Fe_2(CO)_8$ moiety, which contains two identical $Fe(CO)_3$ groups linked to one another by two carbonyl bridges and an Fe,Fe bond. It can be formally derived from the molecular structure of $Fe_2(CO)_9$ by the replacement of one of the three carbonyl bridges with the $Fe(CO)_4$ group. This configuration is in accordance with the observed diamagnetism of $Fe_3(CO)_{12}$.

179

The lengths of the metal–metal bonds, $Fe_1-Fe_3 = 0.2678$, $Fe_1-Fe_2 = 0.2668$ and $Fe_2-Fe_3 = 0.2560$ nm, illustrate the unsymmetrical nature of the triangular arrangement, two atoms being equivalent and the third unique. The Mössbauer spectrum can readily be reconciled with this structure and, indeed, the essential correctness of the above description is supported by the magnitude of the isomer shift for the quadrupole split peaks of the two equivalent iron atoms. This is $+0.280 \pm 0.043$ mm sec^{-1} and is virtually identical with that found in $Fe_2(CO)_9$.

Two other important conclusions have emerged from the stereochemical analysis of $Fe_3(CO)_{12}$ carried out by Wei and Dahl. Firstly, the two Fe,C bond lengths for one bridging carbonyl group are 0.235 and 0.218 nm; for the other, they are 0.224 and 0.208 nm. This illustrates the unsymmetrical nature of the bridging carbonyl groups. Secondly, an explanation has been provided for the unexpectedly simple infrared spectrum of $Fe_3(CO)_{12}$ in solution. Only two bands are observed in the frequency region where terminal carbonyl groups are known to absorb, although nine infrared active stretching modes would be expected on the basis of the structure of C_{2v} symmetry illustrated in Fig. 19.5. This strongly suggests that the molecular state in solution must be different from that in the solid state. It has therefore been proposed that on dissolution, an intramolecular rearrangement of $Fe_3(CO)_{12}$ occurs to give a molecule (of C_{3v} symmetry) consisting of three identical $Fe(CO)_3$ groups located at the vertices of an equilateral triangle and connected in pairs to one another by a bridging carbonyl group and an Fe,Fe bond (Fig. 19.5). The higher symmetry of this species would greatly reduce the number of infrared active bands associated with terminal CO groups compared with those for the molecule in the solid state.

The structures proposed for $Fe_3(CO)_{12}$ have their stereo-chemical counterparts in the two solid state isomeric forms of $Rh_3(C_5H_5)_3(CO)_3$ which are known to exist.[36]

The equivalence of all three iron atoms in $Fe_3(CO)_{12}$ solutions should be demonstrable by Mössbauer spectroscopy. Unfortunately, the low solubility of this in inert solvents has so far made it impossible to measure the spectrum in solution.

In conclusion, it is interesting to note that the molecular configuration of either form of $Fe_3(CO)_{12}$ differs from the structure common to $Os_3(CO)_{12}$ and $Ru_3(CO)_{12}$, containing an equilateral arrangement of three metal atoms, each bonded to four carbonyl groups and linked only by metal–metal bonds.

Bonding in Iron Carbonyls

In carbon monoxide itself there are three occupied σ-orbitals, comprising a σ-bond between the two atoms and two lone pairs of electrons, one largely localized on carbon and the other on oxygen. The lone pairs can be regarded as situated in hybrid orbitals derived from $2s$ and $2p_x$ atomic orbitals. There are also two degenerate π-orbitals, formed by overlap between pairs of p_y and p_z atomic orbitals. These π-orbitals contain four electrons which add considerably to the bond strength.

The electronegativity of oxygen is greater than that of carbon and so it is the lone pair located on carbon which is donated to the metal in carbonyl formation. The electronegativity difference results in some asymmetry in the electron density in the molecule because the σ- and π-bonding electrons are drawn closer to oxygen than to carbon. However, carbon monoxide has a very low dipole moment (~ 0.1 D) and this suggests that the asymmetry is almost exactly offset by a displacement of electrons in the opposite direction. For this reason it is believed that the lone pair on carbon is directed strongly away from the C,O bond. This pair has only slight donor properties, for example, carbon monoxide is not a base and forms only weak complexes with strong electron acceptors such as the boron halides.

The unfilled molecular orbitals in carbon monoxide are two degenerate π-antibonding and one σ-antibonding orbitals. The π-orbitals have the lower energy and are the ones to be filled when carbon monoxide is bonded to a metal possessing d-electrons.

In a metal carbonyl, the lone pair on carbon forms a σ-bond by overlap with an empty orbital on the metal. This by itself would form only a weak bond, not of sufficient strength to account for the known stability of metal carbonyls. Additional bonding results from the overlap possible between an occupied metal orbital of appropriate symmetry and the anti-bonding π-orbitals of the liquid. This is a π-bond and its formation results in a transference of electronic charge from metal to ligand. The accumulation of excess negative charge by the metal, which would be the consequence of σ-bonding on its own, is thereby avoided. π-bonding facilitates the formation of a strong σ-bond and, conversely, the transfer of negative charge to the metal by σ-bonding renders the carbon monoxide ligand positive, thus stimulating more extensive π-bonding. These two effects produce a synergic interaction, in which the resulting combination of σ- and π-bonding is much greater than the bond strength which would be produced by the sum of the effects separately.

It is now possible to account for the formation of metal carbonyls although carbon monoxide has virtually no basic properties. No synergic interaction can occur between a proton and the carbon monoxide molecule because there are no electrons which could participate in π-bonding.

In $Fe(CO)_5$, the trigonal bipyramidal structure results from σ-bond formation involving the $3d_{z^2}$, $4s$- and $4p$-orbitals on the iron atom (Fig. 19.6). The eight valence electrons of iron occupy the d_{xy}, d_{xz}, d_{yz} and $d_{x^2-y^2}$ orbitals and some of these can participate in π-bonding. For example, overlap is possible between a filled t_{2g} orbital with an empty antibonding orbital on carbon monoxide (Fig. 19.6).

In $Fe_2(CO)_9$, the environment of each iron atom is approximately octahedral and the metal may be regarded as using both e_g d-orbitals, as well as $4s$- and $4p$-orbitals, to form six σ-bonds to the carbonyl ligands.[37] On the basis that each terminal carbonyl group

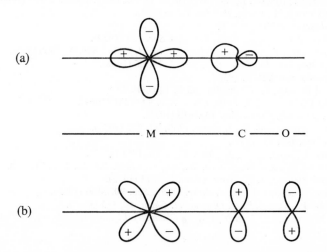

FIG. 19.6. Bonding orbitals in metal carbonyls: (a) overlap to form a σ-bond, and (b) overlap to form a π-bond.

contributes two electrons and each bridging molecule one, each iron atom acquires a share in nine electrons from its ligands. The iron atom provides three of its valence electrons for σ-bonding with the bridging carbonyls and the remaining five electrons are distributed in the three t_{2g} orbitals. One of these can overlap strongly with the corresponding orbital of the other iron atom to form a metal–metal bond: the two other t_{2g} orbitals interact to a much lesser extent with their counterparts of similar symmetry. The metal–metal bond comprises one electron from each iron atom and completes the pairing-up of iron valence electrons. This structural model therefore accounts for the diamagnetism of $Fe_2(CO)_9$. Alternatively, since diamagnetism requires only a weak coupling of unpaired spins, it could be the consequence of multicentre Fe–C–Fe bonding through the bridging carbonyl groups.

References

1. MOND, L., *J. Chem. Soc.* 1090 (1891).
2. BERTHELOT, *Compt. Rend.* **112**, 1343 (1891).
3. EWENS, R. V. G., and LISTER, M. W., *Trans. Faraday Soc.* **35**, 681 (1939).
4. DAVIS, M. I., and HANSON, H. P., *J. Phys. Chem.* **69**, 3405 (1965).
5. DAVIS, M. I., and HANSON, H. P., *J. Phys. Chem.* **71**, 775 (1967).
6. HANSON, A. W., *Acta Cryst.* **15**, 930 (1962).
7. DONOHUE, J., and CARON, A., *Acta Cryst.* **17**, 663 (1964).
8. DONOHUE, J., and CARON, A., *J. Phys. Chem.* **71**, 777 (1967).
9. BERGMANN, E., and ENGEL, L., *Z. Physik. Chem.* B **13**, 232 (1931).
10. GRAFFUNDER, W., and HEYMANN, E., *Z. Physik. Chem.* B **15**, 377 (1932).
11. HORROCKS, W. D., and DiCARLO, E. N., *J. Phys. Chem.* **66**, 186 (1962).
12. O'DWYER, M. F., *J. Mol. Spectry.* **2**, 144 (1958).
13. SHELINE, R. K., and PITZER, K. S., *J Amer. Chem. Soc.* **72**, 1107 (1950).
14. KING, F. T., and LIPPINCOTT, E. R., *J. Amer. Chem. Soc.* **78**, 4192 (1956).
15. STAMMREICH, H., SALA, O., and TAVANES, Y., *J. Chem. Phys.* **30**, 856 (1959).
16. EDGELL, W. F., WILSON, W. E., and SUMMITT, R., *Spectrochim. Acta* **19**, 863 (1963).
17. COTTON, F. A., DANTI, A., and WAUGH, J. S., *J. Chem. Phys.* **29**, 1427 (1958).
18. BRAMLEY, R., FIGGIS, B. N., and NYHOLM, R. S., *Trans. Faraday Soc.* 1893 (1962).
19. EPSTEIN, L. M., *J. Chem. Phys.* **36**, 2731 (1962).
20. HERBER, R. H., KINGSTON, W. R., and WERTHEIM, G. K., *Inorg. Chem.* **2**, 153 (1963).
21. SPEYER, E., and WOLF, H., *Chem. Ber.* **60**, 1424 (1927).
22. *Inorganic Syntheses*, vol. VIII, p. 178.
23. POWELL, H. M., and EWENS, R. V. G., *J. Chem. Soc.* 286 (1939).
24. GIBB, T. C., GREATREX, R., and GREENWOOD, N. N., *J. Chem. Soc.* A, 890 (1968).
25. DEWAR, J., and JONES, H. O., *Proc. Roy. Soc.* A **79**, 66 (1966).
26. *Inorganic Syntheses*, vol. VII, p. 193.
27. HIEBER, W., and BECKER, E., *Chem. Ber.* **63** B, 1405 (1930).
28. WEI, C. H., and DAHL, L. F., *J. Amer. Chem. Soc.* **91**, 1351 (1969).
29. SHELINE, R. K., *J. Amer. Chem. Soc.* **73**, 1615 (1951).
30. DAHL, L. F., and RUNDLE, R. E., *J. Chem. Phys.* **27**, 323 (1957).
31. MILLS, O. S., *Chem. Ind.* 73 (1957).
32. DAHL, L. F., and RUNDLE, R. E., *J. Chem. Phys.* **26**, 1751 (1957).
33. FLUCK, E., KERLER, W., and NEUWIRTH, W., *Angew. Chem. Int. Ed.* **2**, 277 (1963).
34. DAHL, L. F., and BLOUNT, J. F., *Inorg. Chem.* **4**, 1373 (1965).
35. DAHM, D. J., and JACOBSON, R. A., *J. Amer. Chem. Soc.* **90**, 5106 (1968).
36. MILLS, O. S., and PAULUS, E. F., *J. Organomet. Chem.* **10**, P 3 (1967).
37. ORGEL, L. E., *Introduction to Transition Metal Chemistry-Ligand-Field Theory*, Methuen, London (1960)

CHAPTER 20

FERROCENE

Introduction

The bonding in coordination compounds is conventionally described in terms of the sharing of an electron pair between two atoms, one, the donor or ligand, which provides the two electrons and the other, the acceptor, which receives the pair into one of its vacant orbitals. The donor atom may form part of an inorganic or an organic molecule or ion but it is considered that the electron pair which constitutes the coordinate bond with a metal atom or ion is originally localized as a 'lone-pair' on the donor atom. This simple concept affords a satisfactory description of bond formation in many coordination compounds.

Recently, it has become clear that, in certain classes of compounds, the bonding from ligands to metal must involve electron pairs which occupy delocalized orbitals in the uncoordinated ligand. Pre-eminent among these compounds are the organometallic complexes involving cyclo-pentadiene and other cyclic olefins, benzene and other arenes, and acetylenes. The electronic structures of such ligands and their complexes are most appropriately described by molecular orbital theory.

The synthesis[1,2] of bis(cyclopentadienyl) iron ('ferrocene') $Fe(C_5H_5)_2$, in 1951, was a crucial stage in the development of chemistry for several reasons. It was quickly recognized that ferrocene has a unique 'sandwich' structure,

in which the iron atom is located between two cyclopentadienyl rings having pronounced aromatic character. The discovery of ferrocene provided the stimulus for the synthesis of many other metallocenes of general type, $M(C_5H_5)_2$, and of mixed complexes, derived from the bis(cyclopentadienyl) compounds by replacement of one C_5H_5 ring by other ligands as, for example, in $(C_5H_5)Mn(CO)_3$. In view of the many cyclopentadienyl complexes which were made within a short period following the discovery of ferrocene, it was natural to seek complexes of other unsaturated rings. It was soon established that molecules like cycloheptatriene and tetraphenyl cyclobutadiene could act as ligands in a similar way to cyclopentadiene. The preparation,[3] in 1956, of bis(benzene) chromium, $Cr(C_6H_6)_2$, showed that aromatic molecules could also form 'sandwich' complexes with metal atoms.

The reactions of ligands in organometallic complexes have been extensively studied and provide valuable insight into the effect of coordination on the properties of the organic molecules. Such complexes are very probably formed as reaction intermediates during the catalytic hydrogenation of unsaturated compounds.

The nature of bonding in a molecule like ferrocene has proved a controversial topic for theoretical chemists and many qualitative and quantitative descriptions have been published. The development of the chemistry of the metallocenes has resulted in fruitful interactions between theory and experiment. Theoretical treatments of the bonding have been of predictive value for experimental work and the preparation of such novel compounds has resulted in important theoretical advances.

Preparation

Ferrocene was prepared by Kealy and Pauson[1] in 1951 by the reaction between cyclopentadienyl magnesium bromide and iron(III) chloride in diethyl ether solution. This was being studied in an attempt to prepare fulvalene, $(C_5H_5)_2$, but the product isolated was ferrocene. Iron(II) chloride is first formed by the reducing action of cyclopentadiene, and then reaction with the Grignard reagent proceeds thus:

$$2C_5H_5MgBr + FeCl_2 \rightarrow Fe(C_5H_5)_2 + MgBr_2 + MgCl_2$$

At about the same time, Miller, Tebboth and Tremaine[2] reported the formation of ferrocene by direct reaction at 300°C between cyclopentadiene vapour and an iron catalyst containing aluminium, potassium or molybdenum oxides. The catalyst soon becomes passive and requires reactivation from time to time.

The alkali metal salts of cyclopentadiene have been used[4,5,6] successfully in the synthesis of ferrocene and other bis(cyclopentadienyl) complexes. The sodium or potassium salt is made by the treatment of cyclopentadiene with the alkali metal in a solvent such as tetrahydrofuran. Subsequent treatment of the solution of the alkali metal cyclopentadienide with iron(III) or (II) chloride dissolved in tetrahydrofuran gives ferrocene. Iron(II) chloride is preferably used, otherwise some of the cyclopentadienide is used up in the reduction of iron(III) to iron(II).

A simplification in experimental procedure which avoids the need to prepare the cyclopentadienide in a separate stage involves the use[7] of a base such as diethylamine as solvent for the cyclopentadiene and the treatment of this solution with anhydrous iron(III) or (II) chloride. Cyclopentadiene is a weak acid which is converted by the amine in small but significant amounts to the cyclopentadienide ion, $C_5H_5^-$. Reaction with the metal chloride is carried out at 0°C under a nitrogen atmosphere, then almost all the solvent is removed by evaporation and the ferrocene isolated by steam distillation. This method has been applied successfully to the synthesis of ferrocene enriched with particular iron isotopes and analogous procedures have been used to make nickelocene and cobaltocene.

Instead of alkali metal salts of cyclopentadiene, other cyclopentadienyl complexes have been used as a source of $C_5H_5^-$ ions in the preparation of ferrocene. For example, bis-(cyclopentadienyl) magnesium, which may be prepared directly from cyclopentadiene and magnesium, is an ionic compound which reacts directly with iron(II) chloride to give ferrocene.

Another variant in preparative reactions is the use of iron pentacarbonyl instead of iron itself or one of its salts. For example, ferrocene can be made by passing a mixture of cyclopentadiene and iron pentacarbonyl vapours through a heated tube. If the temperature is between 100° and 180°C, the dinuclear mixed complex, $[C_5H_5Fe(CO)_2]_2$ is obtained. This serves as a source of many monocyclopentadienyl derivatives of iron because one or more of the CO groups can be replaced by other ligands. On heating the dinuclear complex to 210°C, it loses CO and ferrocene is formed.

Reactions

Ferrocene is a diamagnetic, orange solid which melts at 173°C and is thermally stable up to 470°C.

It is converted by oxidizing agents such as bromine and dilute nitric acid, to the ferricinium ion,[8] $Fe(C_5H_5)_2^+$. Solutions containing this are blue and it forms sparingly soluble salts with large anions like tetraphenylborate, chloroplatinate and silicotungstate. Ferricinium picrate has an effective magnetic moment, $\mu_{eff} = 2.26$ B.M., and this corresponds with the presence of one unpaired electron in the cation.

The chief interest in the chemistry of ferrocene lies in the substitution reactions which the cyclopentadienyl rings can undergo. Pre-eminent among these is the Friedel–Crafts reaction in which hydrogen is replaced by an acyl or alkyl group.

For example, ferrocene in carbon disulphide solution reacts with acetyl chloride and aluminium chloride to give a mixture of mono and diacetylferrocenes. In place of aluminium chloride, other Lewis acids such a $SnCl_4$, BF_3 and HF may be used. The reaction demonstrates the aromatic nature of the hydrocarbon rings in ferrocene. A remarkable feature is the high reactivity of ferrocene relative to other aromatic compounds. Thus it has been estimated[9] that the rate of acetylation of ferrocene is 3.3×10^6 times that of benzene itself.

Acylation of ferrocene with equimolar amounts of reagents yields mainly the monoacyl derivative. In the presence of excess $AlCl_3$, diacylferrocenes are formed in significant quantities when equimolar amounts of ferrocene and the acyl halide are used and the di-substituted derivatives are obtained in high yield when ferrocene is treated with excess acyl halide.

The diacylferrocenes which have been isolated are almost invariably the symmetrically substituted derivatives with one acyl group on each ring. For example, from acetyl chloride, 1,1-diacetylferrocene(II) is obtained. The original

assignment of this structure was based[10] on its oxidation by hypoiodite to a ferrocene dicarboxylic acid, for which the first and second dissociation constants are: $pK_1 = 3.1 \times 10^{-7}$ and $pK_2 = 2.7 \times 10^{-8}$. The difference between these values is very small, indicating that the carboxyl groups interact to a very limited extent and hence that they are far apart (that is, on different rings). This conclusion was reaffirmed by the subsequent observation that hydrogenation of diacetylferrocene gives ethylcyclopentane but no cyclopentane or diethylcyclopentane.

The electrophilic substitution of ferrocene appears to proceed[9] by the initial formation of a charge-transfer complex (III) between the electrophile, E, and the orbital electrons of the metal. Support for this mechanism comes from the discovery that ferrocene can be protonated, for example, in solution in boron trifluoride hydrate, by direct bonding between hydrogen and the iron atom.[11,12] That protonation occurs at the metal is shown by the very high field proton magnetic resonance observed for such solutions. This example provides clear evidence for the accessibility of electrons for bonding, the electrons concerned occupying the highest filled (E_{2g}) levels. These are essentially non-bonding in the ferrocene molecule itself (see below). The intermediate charge-transfer complex rearranges to (IV), resulting in a partial transfer of charge to the ring undergoing substitution. Finally, loss of the proton gives the substituted ferrocene (V).

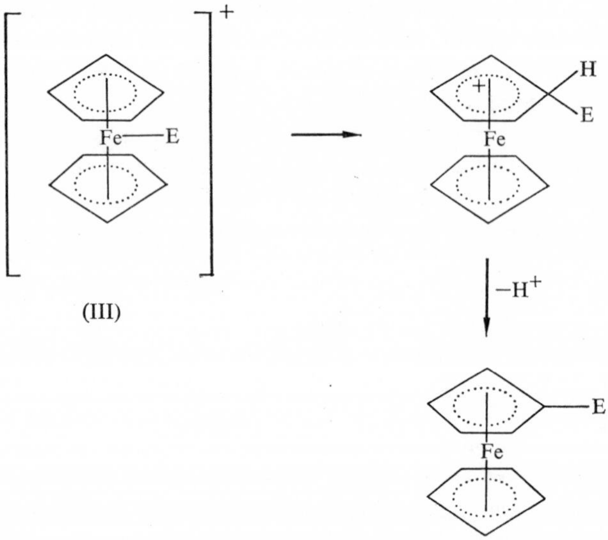

(III)   (IV)   (V)

The great reactivity of ferrocene is again illustrated by its Vilsmeier formylation, i.e. reaction with N-methylformanilide and phosphorus oxochloride to give ferrocenecarboxy-

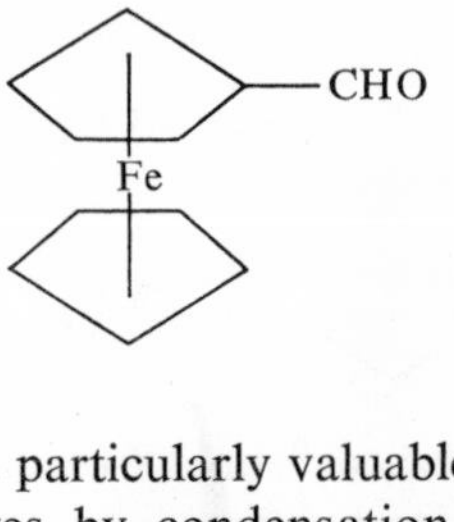

(VI)

aldehyde (VI). This compound is a particularly valuable starting material for the preparation of many ferrocene derivatives by condensation reactions involving the aldehyde grouping.

186

The alkylation of ferrocene with alkyl halides or olefins occurs under Friedel–Crafts conditions. In general, yields are small and the products are mixtures of mono- and poly-alkyl derivatives.

Ferrocene reacts with formaldehyde in the presence of HF or concentrated H_2SO_4 to give a dinuclear compound 1,2-diferrocenylethane (VII).

Aminomethylation of ferrocene occurs with formaldehyde and dimethylamine in acetic acid under the conditions of the Mannich reaction. The quaternary ammonium salt (VIII) is the starting material for many other ferrocene derivatives.

Arylferrocenes are most readily obtained by the reaction of ferrocene with aryl diazonium salts in aqueous acetic acid, acetone solutions or chlorohydrocarbon solvents. The primary products are mono-substituted derivatives.

Ferrocene does not show the reactions typical of polyolefins. It does not react with maleic anhydride and cannot be hydrogenated using a platinum catalyst. Rapid cleavage of the molecule into metallic iron and cyclopentadiene is effected by using lithium in di-ethylamine[13] as reducing agent.

The ease of oxidation of ferrocene to ferricinium ions precludes the direct nitration of the cyclopentadienyl rings. Amino and nitro groups can be introduced into the rings by using lithium-substituted ferrocenes as intermediates. Ferrocene in solution in anhydrous ether reacts with n-butyllithium[14] to give a mixture of mono- and dilithium ferrocenes. Treatment of these with CO_2, followed by hydrolysis with water and HCl, produces the corresponding mono and 1,1'-dicarboxylic acids. Monolithium ferrocene reacts with meth-oxyamine[15] to give ferrocenylamine, $FeC_{10}H_9NH_2$: with N_2O_4 in ether solution at $-70°C$ it gives nitroferrocene.[16]

The rings in ferrocene can also be mercurated by reaction with mercury(II) chloride. This reaction, when carried out in either ether/ethanol or benzene/ethanol mixtures, produces a mixture of chloromercuriferrocene (IX) and 1,1′-dichloromercuriferrocene (X).

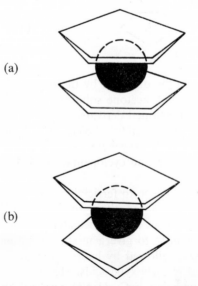

$$\text{(IX)} \qquad\qquad\qquad\qquad \text{(X)}$$

Chloromercuriferrocenes react with iodine or bromine in xylene to give mono- or di-substituted iodo- or bromoferrocenes. This is an important reaction because direct halo-genation of ferrocene rings is not possible due to oxidation to ferricinium compounds.

Structure

In the gaseous state, vapour density measurements[17] have shown that ferrocene consists of monomeric, undissociated molecules. The solid is diamagnetic and in solution, the dipole moment of ferrocene is zero.[8] Its infrared spectrum shows a single band, at $3080\,\mathrm{cm^{-1}}$ in the region where C–H stretching frequencies occur and its N.M.R. spectrum shows a single, sharp resonance peak[18,19]. These properties demonstrate that all the hydrogens in ferrocene are structurally and magnetically equivalent. This equivalence together with the high symmetry of the molecule, indicated by the absence of a dipole moment, are in accordance with the 'sandwich' structure, in which the iron atom is centred between two parallel cyclopentadienyl rings, proposed by Wilkinson, Rosenblum, Whiting and Woodward.[8]

The two rings may be in an eclipsed or a staggered configuration (Fig. 20.1a and b respectively). The first electron diffraction study on the vapour showed[20] that, at 400°C, the eclipsed form predominates and the cyclopentadienyl rings are able to rotate freely

(a)

(b)

FIG. 20.1. Eclipsed (a) and staggered (b) arrangements of the cyclopentadienyl rings in ferrocene molecules.

within the molecule. This form has D_{5h} symmetry. The energy barrier to rotation of the rings about the five-fold symmetry axis was estimated as 4–8 kJ mol⁻¹. With such a small barrier, the possibility of such rotation occurring in condensed states must also be considered.

The more detailed electron diffraction study carried out by Bohn and Haaland[21] has confirmed that the eclipsed configuration exists in the vapour at 140°C. The principal molecular parameters are: C,C = 0·1431 ± 0·0005 nm; Fe,C = 0·2058 ± 0·0005 nm; and C,H = 0·1122 ± 0·002 nm. The inter-ring distance is 0·3319 ± 0·0015 nm. There appears to be some deformation of each ligand by bending of the C–H bonds about 5° out of the plane of the C₅ ring towards the iron atom.† This converts the D_{5h} local symmetry of the planar C_5H_5 ring into C_{5v} symmetry of the bent configuration. This molecular structure is illustrated in Fig. 20.2. The height of the barrier to internal rotation is given by these authors as about 4·6 kJ mol⁻¹.

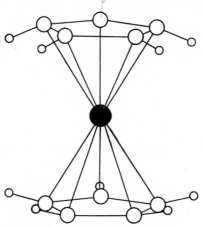

FIG. 20.2. The structure of the ferrocene molecule as determined by electron diffraction (reproduced with permission from Bohn, R. K., and Haaland, A., *J. Organomet. Chem.* **5**, 470 (1966)).

In benzene solution, the dipole moments[22] of mono- and disubstituted ferrocenes are in close agreement with values calculated on the assumption that free rotation of each ring with respect to the other can occur.

In the crystalline state, however, it is clearly established, by X-ray diffraction studies, that it is the staggered form of the ferrocene molecule which predominates at ordinary temperature.[23,24] Ferrocene belongs to the space group P_{2_1}/a. The unit cell is monoclinic with dimensions: a = 1·050 nm; b = 0·763 nm; c = 0·595 nm; and β = 121°. It contains two formula units of $C_{10}H_{10}Fe$, each being centrosymmetric about the iron atom. On the assumption that there is no significant departure from D_{5d} molecular symmetry, the inter-atomic distances have been calculated as: C,C = 0·1403 nm; Fe,C = 0·2045 nm. The electron density map shows there is an accumulation of electron density between the atoms of the ring and a corresponding deficit at the atoms themselves. This could result from torsional vibration of the rings about the five-fold symmetry axis. However, free rotation, either of the whole molecule or of the cyclopentadienyl rings within the molecule, is not possible. In view of this difference from the behaviour of isolated ferrocene molecules,

† Haaland and Nilsson (*Acta Chem. Sound.* **22**, 2653 (1968)) have stated that more recent work on ferrocene suggests there is no real bending of the C–H bonds towards the metal.

the restriction of rotation and the stabilization of D_{5d} symmetry in the crystal state must be primarily due to intermolecular forces.

A λ-point transition centred at $-109 \cdot 3°C$ occurs[25] in the heat capacity curve of ferrocene and it has been adduced as evidence for structural changes in the crystal. Values of C_p, the heat capacity at constant pressure, increase from about 84 $JK^{-1} mol^{-1}$ at $-153°C$ to a maximum of 1820 $JK^{-1} mol^{-1}$ at $-109 \cdot 3°C$. Thereafter, C_p falls to a first minimum of $ca.$ 130 $JK^{-1} mol^{-1}$ and then there is a subsidiary maximum at $-109 \cdot 3°C$ before returning to normal values at $-73°C$

X-ray diffraction studies over the range -178 to $+17°C$ have shown that the expansion coefficients of the lattice are anisotropic. The consequence of this on the molecular packing in the crystal must be considered. The three types of interaction present between two cyclopentadienyl rings on adjacent ferrocene molecules are illustrated in Fig. 20.3 (i), (ii) and

(i)

(ii)

(iii)

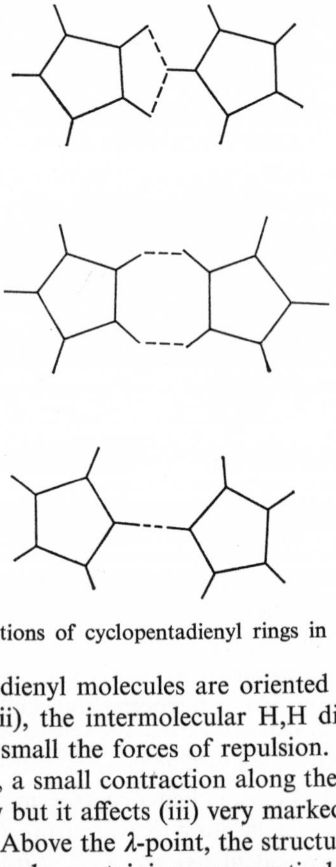

FIG. 20.3. Possible orientations of cyclopentadienyl rings in solid ferrocene.

(iii). The planes of the cyclopentadienyl molecules are oriented as shown with respect to the molecular c axis. In (i) and (ii), the intermolecular H,H distances are large enough ($>$ or $= 0 \cdot 25$ nm) to render very small the forces of repulsion. In (iii), the H,H distance is about $0 \cdot 215$ nm. For (i) and (ii), a small contraction along the c-axis produces no significant change in interaction energy but it affects (iii) very markedly and causes a relatively large increase in potential energy. Above the λ-point, the structure of ferrocene is believed to be partially disordered, for example, containing comparatively large regions of eclipsed molecules surrounding or being surrounded by large domains of the predominating species, the staggered molecules. This structure becomes unstable at low temperatures with respect to a completely ordered lattice of staggered molecules, mainly because of the effect of lattice contraction on the energy of interactions of type (iii). The changeover from one structure to the other may be responsible for the λ-point transition at $-109 \cdot 3°C$. Confirmation of structural changes over the temperature range where heat capacity anomalies are observed has been found in the N.M.R. spectrum of crystalline ferrocene.

N.M.R. Spectrum

The variation with temperature of the width of the single resonance line has been interpreted[26,27] in terms of the reorientation of the cyclopentadienyl rings about their five-fold symmetry axes. This resonance shows a progressive narrowing when the temperature of measurement is decreased from $-48°$ to $-158°C$. This has been attributed to the conversion of molecules in the eclipsed to the staggered configuration stable at low temperatures. The interactions between protons in the cyclopentadienyl rings of the ferrocene molecule are larger (because the interproton distance is shorter) in the eclipsed than in the staggered configuration. Above $-158°C$, the resonance line width increases as contributions from the protons of eclipsed molecules become significant.

Vibrational Spectrum

The vibrational spectrum of ferrocene[28] is consistent with a molecular structure of D_{5d} symmetry. The number of normal vibrations expected for any non-linear molecule is equal to $(3N–6)$ where N = number of atoms. For ferrocene, $N = 21$ and the total number of vibrations = 57. Each C_5H_5 ring has 24 vibrations: for two rings there are 48 vibrations and the remaining 9 vibrations are associated with the 'skeleton' of the sandwich molecule.

The cyclopentadienyl anion, $C_5H_5^-$, has a planar pentagonal structure with D_{5h} symmetry. There is a five-fold axis perpendicular to the ring through its centre, five two-fold axes perpendicular to this axis and a horizontal symmetry plane containing the ring.

In a sandwich structure with covalent bonding between the metal and $C_5H_5^-$ ions, the local symmetry of each ring is reduced to C_{5v}, that of a pentagonal bipyramid. The molecule as a whole has D_{5d} symmetry. It has the following symmetry elements: one C_5 axis extending through the iron atom and centres of mass of the two cyclopentadienyl rings; one S_{10} axis collinear with the C_5 axis; five reflection planes (σ_v) each containing the C_5 axis and each bisecting the angle between two planes; and a centre of inversion coincident with the iron atom. Table I give the selection rules for a sandwich structure of this symmetry.

Raman-active fundamentals are in the species A_{1g}, E_{1g} and E_{2g} and infrared-active fundamentals are in the species A_{2u} and E_{1u} and there should be 15 Raman and 10 infrared fundamental frequencies observable in the vibrational spectrum. The assignment of frequencies corresponding to ligand vibrations is facilitated by comparison to similar modes in benzene and by measurements of the spectrum of the deuterated analogue, $Fe(C_5D_5)_2$.

TABLE I

Selection Rules for Ferrocene

Species	Number of frequencies	Activity
A_{1g}	4	R
A_{1u}	2	Inactive
A_{2g}	1	Inactive
A_{2u}	4	i.r.
E_{1g}	5	R
E_{1u}	6	i.r.
E_{2g}	6	R
E_{2u}	6	Inactive

R = Raman active; i.r. = infrared active

Frequencies due to ring–metal–ring vibrations have been assigned on the assumption they would be roughly analogous to those observed for linear XY_2 molecules.

Examination of the Raman and infrared spectra of ferrocene shows that many frequencies are coincident or nearly coincident in both, despite the centre of symmetry possessed by the D_{5d} model which should result in the rule of mutual exclusion being obeyed. That coincidences do exist has been interpreted to mean there is little interaction between the symmetric and antisymmetric modes of vibration involving displacements associated with the two cyclopentadienyl rings. These coincidences are accordingly identified with the C–H and C–C modes of vibrations.

The frequency assignments[29] made by Stammreich are given in Table II. The normal vibrations of the irreducible representation E_{2u} were found from the spectra of single crystals where they are infrared active due to crystal forces.

The band system observed in the infrared spectrum around 1700 cm^{-1} has been interpreted in different ways. These bands do not appear in the Raman spectrum and cannot be regarded as due to fundamental vibrational modes. Also there is no structural feature associated with D_{5d} symmetry which is expected to give rise to bands in this region. They have been related[30] to the restricted rotation of the cyclopentadienyl rings with respect to one another. However, they are also prominent in the infrared spectrum of crystalline

TABLE II

Frequency Assignments for the Ferrocene Molecule of D_{5d} Symmetry

In-phase vibrations			Out-of-phase vibrations	
Irreducible representation	Frequency (cm^{-1})	Approx. type of vibration	Irreducible representation	Frequency (cm^{-1})
$A_{1g}\,\nu_1$	3110 (3100)	CH stretch	$A_{2u}\,\nu_8$	3086
ν_3	1390 (1105)	Ring breathing	ν_{10}	1408
ν_4	306 (301)	M-ring stretch	ν_{11}	478
$A_{2u}\,\nu_9$	1104	CH bend ($\perp$)	$A_{1g}\,\nu_2$	1105 (812)
$A_{2g}\,\nu_7$	(1249)*	CH bend ($\parallel$)	$A_{1u}\,\nu_5$	(1253)*
		Torsion	ν_6	—
$E_{1g}\,\nu_{14}$	818 (835)	CH bend ($\perp$)	$E_{1u}\,\nu_{19}$	814
ν_{16}	390 (390)	Ring tilt	ν_{21}	490
		Ring-M-ring bend	ν_{22}	170
$E_{1u}\,\nu_{17}$	3086	CH stretch	$E_{1g}\,\nu_{12}$	3089 (3085)
ν_{18}	1004	CH bend ($\parallel$)	ν_{13}	998 (999)
ν_{20}	1408	CC stretch	ν_{15}	1412 (1412)
$E_{2g}\,\nu_{23}$	3045 (3070)	CH stretch	$E_{2u}\,\nu_{29}$	(3035)*
ν_{24}	1361 (1175)	CH bend ($\parallel$)	ν_{30}	(1351)*
ν_{26}	1527 (1356)	CC stretch	ν_{32}	—
ν_{27}	1054 (892)	Ring distortion ($\parallel$)	ν_{33}	(1054)*
$E_{2u}\,\nu_{31}$	(1188)*	CH bend ($\perp$)	$E_{2g}\,\nu_{25}$	1184 (1059)
ν_{34}	(567)*	Ring distortion ($\perp$)	ν_{28}	591 (600)

($\parallel$) and ($\perp$) denote vibrations parallel and perpendicular respectively to the z-axis of the molecule (this passes through the metal atom and is perpendicular to the cyclopentadienyl rings).

Figures given thus: 3110 are assignments of Stammreich reported by Fritz.[29]

Figures in parentheses: assignments of Long and Huege.[33]

Figures given thus: (1188)* are calculated values since the vibrations concerned are inactive.

ferrocene[31] and it seems unlikely that hindered rotation would be significant here. It appears more likely that they are overtone and combination bands associated with C–H bending modes.

It has been pointed out[32] that the assignments made by Lippincott and Nelson are based on the assumption that the rings were aromatic and that parallels with benzene might therefore be expected. The success in interpreting the vibrational spectrum of ferrocene supports the validity of this assumption. It has been found that the infrared spectrum is a valuable criterion in establishing the structure of cyclopentadienyl metal complexes in general for a close correspondence between the spectrum of another complex species and that of ferrocene strongly suggests they both have similar sandwich structures. This kind of experimental approach has been used to establish the ferrocene-like structure of ions such as $[W(C_5H_5)_2]^{2+}$, $[Mo(C_5H_5)_2]^{2+}$, $[V(C_5H_5)_2]^{2+}$, $[Ti(C_5H_5)_2]^{2+}$ and $[Nb(C_5H_5)_2]^{3+}$. Conversely, when the spectrum is more complex than that of ferrocene, this is indicative of a different structure and kind of bond within the complex. For example, the spectrum of $Hg(C_5H_5)_2$ shows additional bands such as multiple C–H stretching frequencies. All C–H bonds are not spectroscopically equivalent, as in ferrocene, because the mercury complex contains σ-bonds between the metal and specific carbon atoms of the cyclopentadienyl rings.

The Raman spectrum of ferrocene[33] in various solvents and in the crystalline state has recently been measured using the helium–neon gas laser as radiation source for excitation of the Raman lines. The advantage of this type of excitation is that it is so much more intense than that from, for example, a mercury lamp. The complete Raman spectrum can be photographed in a small fraction of a second compared with the much longer exposure times, sometimes as great as several hours, required when using conventional sources. The intense polarized band at 1390 cm^{-1} observed by Stammreich (see ref. 30) was not observed by Long and Huege.[33] These authors prefer the original assignment of $\nu_3 = 1105$ cm^{-1} given by Lippincott and Nelson. ν_2 had not been previously seen owing to solvent masking but now has been observed directly at 812 cm^{-1}. Although the E_{1g} assignments by Long and Huege are substantially the same as those of Stammreich, there are considerable differences in assignments made for the E_{2g} vibrations.

Mössbauer Spectrum

Several investigations of the Mössbauer spectrum of ferrocene have been carried out. This spectrum is characterized[34] by a quadrupole splitting of 0·236 cm sec^{-1} and an isomer shift of 0·060 cm sec^{-1} at 25°C (values of these quantities at $-195°C$ are 0·237 and 0·068 cm sec^{-1} respectively). The spectrum in the absence of an applied magnetic field is shown in Fig. 20.4, the quadrupole splitting being caused by the interaction of the nuclear quadrupole moment of ^{57}Fe with the gradient of the electric field at the nucleus. The energy level diagram for ^{57}Fe shows the splitting of the first excited state caused by the electric field gradient, V_{zz}, which is symmetric about the z axis of the ferrocene molecule.

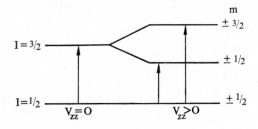

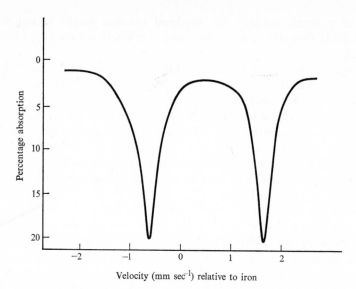

Velocity (mm sec^{-1}) relative to iron

FIG. 20.4. Mössbauer absorption spectrum of ferrocene at $-268 \cdot 8°C$

The main contribution to V_{zz} is believed to be from the $3d$ electrons. Evidence to support this view is that quadrupole splitting vanishes in ferricinium compounds. The oxidation of ferrocene to ferricinium ion involves the removal of an electron from the highest filled energy level in the molecule. Anticipating the molecular orbital description of the bonding given below, this is the M.O. of symmetry E_{2g}, which is very largely derived from $3d$ metal orbitals. The removal of d electron density, while the $4p$ density remains unchanged, accompanied as it is by the collapse of quadrupole splitting, is consistent with the view that the $3d$ electrons make the largest contribution to the quadrupole splittings at the iron nucleus.

Bonding

The remarkable 'sandwich' structure of ferrocene and other cyclopentadienyl complexes has aroused great interest and resulted in many theoretical treatments of the bonding.

The formulae of many transition metal complexes had, for some years prior to the discovery of ferrocene, been rationalized by use of the Effective Atomic Number rule, according to which the metal gains enough electrons from the ligands to complete its valence shell of 18 electrons and reach the electronic configuration of a noble gas, and it was natural to attempt also to describe the bonding in ferrocene by this concept. E. O. Fischer and co-workers[35] suggested that the electron shells of the iron(II) ion accept six electrons from each cyclopentadienyl ion. These electrons are situated in pairs at the corners of an equilateral triangle in each ring and on bonding to iron are donated to the empty d^2sp^3 hybrid orbitals of the metal. When described in this way, ferrocene strongly resembles the so-called penetration complexes such as $Fe(CN)_6{}^{4-}$.

Alternatively, the metal can be regarded as in a zero oxidation state (thus having 8 electrons in its valence shell) and bound to 2 cyclopentadienyl radicals, each contributing 5 electrons to the bonding. Again the total number of electrons in the valence shell of the bound metal totals 18 and this has the krypton configuration. This description differs only in a formal manner from the preceding one but it is a useful one to bear in mind when considering some of the properties of ferrocene, for example, the interchangeability of carbon monoxide and cyclopentadiene as ligands resulting in the formation of mixed complexes.

The properties of ferrocene have established that the cyclopentadienyl rings possess aromatic character. This suggests delocalized electrons should be available within the ring and it is difficult to see how they can simultaneously be involved in bonding with the metal to the extent expected in a penetration complex.

The synthesis of the bis-cyclopentadienyls of other first row transition metals, such as V, Cr, Co and Ni, quickly followed the discovery of ferrocene. From this, it is clear that the attainment of a noble gas configuration by the metal is not a necessary condition for the formation of such complexes. Indeed, it has been recognized for some time that the Effective Atomic Number rule is generally only of limited utility in describing the electronic structure of complexes.

Several theoretical treatments of the bonding in ferrocene based on molecular orbital theory have been published. The essence of these is the classification of the orbitals of the metal and the ligands according to their symmetry properties in the point group, D_{5d}, followed by the combination of orbitals of the same symmetry type to form molecular orbitals which encompass the whole molecule.

Classification of Orbitals of the Metal

The $3d$, $4s$ and $4p$ metal orbitals (Fig. 20.5) are classified in the D_{5d} point group by the representations: $A_{1g}(3d_{z^2}, 4s)$; $A_{2u}(4p_z)$; $E_{1g}(3d_{xz}, 3d_{yz})$; $E_{2g}(3d_{xy}, 3d_{x^2-y^2})$ and $E_{1u}(4p_x, 4p_y)$.

The d-orbitals are distinguished by the components of angular momentum, $m\left(\dfrac{h}{2\pi}\right)$, which they have about a given axis ($m = 0, \pm 1$, and ± 2). Their wave functions, expressed as the product of a radial function, $R(r)$, and an angular function involving polar co-ordinates θ and ϕ, are:

$$d_{E_{2g}} = (15/32\pi)^{\frac{1}{2}} \sin^2\theta\, e^{\pm 2i\phi}\, R(r): \quad m = \pm 2$$

$$d_{E_{1g}} = (15/8\pi)^{\frac{1}{2}} \sin\theta\, \cos\theta\, e^{\pm i\phi}\, R(r): \quad m = \pm 1$$

$$d_{A_{1g}} = (5/16\pi)^{\frac{1}{2}} (3\cos^2\theta - 1)\, R(r): \quad m = 0$$

These orbitals are *gerade* and thus given the subscript g because no change in the sign of the wave-function results from the symmetry operation of inversion in the atomic nucleus.

The effect of a rotation, α, about the z-axis (perpendicular to the planes of the cyclopentadienyl rings passing through their centres and the iron atom) clarifies the distinction of these orbitals from each other. This rotation is equivalent to constructing a new set of orbitals whose values at the point (r, θ, ϕ) are the same as those of the original at the point $(r, \theta, \phi + \alpha)$. When $m = 0$, no change results from the rotation because the orbital is independent of ϕ. This orbital is designated A_{1g}. When $m = \pm 1$, the orbitals resulting from rotation contain the additional factors $e^{\pm i\alpha}$: when $m = \pm 2$, the orbitals contain

195

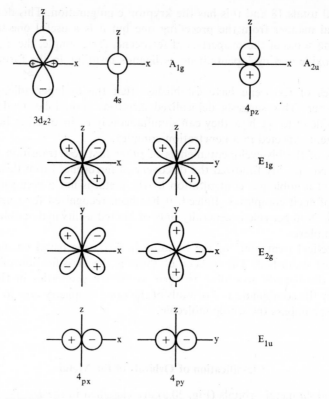

FIG. 20.5. Metal atom orbitals and their representations in the D_{5d} symmetry group.

the additional factors $e^{\pm 2i\alpha}$. These constitute conjugate pairs of orbitals, designated respectively E_{1g} and E_{2g}.

The spherically symmetrical 4s-orbital is *gerade* and remains invariant under the rotation and transforms according to A_{1g}. The 4p-orbitals are *ungerade* because the wave function changes sign on inversion. The p_z-orbital ($m = 0$) is invariant under any rotation about the z-axis and p_x and p_y (for which $m = \pm 1$) are multiplied by $e^{\pm i\alpha}$. These orbitals are designated A_{2u} and E_{1u} respectively.

Classification of Orbitals of the Cyclopentadienyl Rings

Firstly, the five p_z atomic orbitals on each ring are combined to give five localized molecular orbitals (Fig. 20.6). These are classified as A_1, E_1 and E_2, according to their symmetry properties with respect to rotation about the molecular axis perpendicular to each ring and passing through its centre. The orbital designated A_1 has no nodal plane perpendicular to the plane of the ring and remains invariant to rotation about the five-fold symmetry axis; E_1 has one nodal plane and E_2 has two nodal planes perpendicular to this plane.

Secondly, these localized molecular orbitals are combined linearly to give a set of 10 orbitals which extend over both rings. These are designated A_{1g}, A_{2u}, E_{1g}, E_{1u}, E_{2g} and

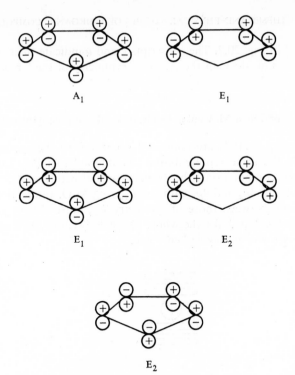

FIG. 20.6. Molecular orbitals of the cyclopentadienyl ring shown in terms of their component p_z atomic orbitals.

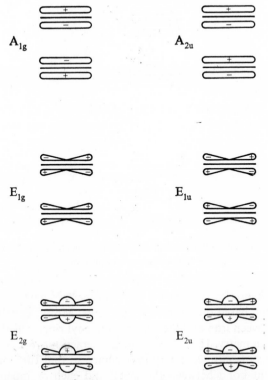

FIG. 20.7. Molecular orbitals for the two cyclopentadienyl rings in ferrocene. Only one member of a pair of degenerate orbitals is shown. The C_5H_5 ring is shown in each case as the central horizontal line.

E_{2u}, and illustrated in Fig. 20.7. The subscripts g and u indicate their symmetry properties with respect to inversion through the centre of symmetry of the ferrocene molecule.

Combined Molecular Orbitals for the Whole Molecule

The metal and ring orbitals are combined according to the symmetry rule that only those belonging to the same representation can interact together. Figure 20.8 illustrates some of the allowed combinations: the $4s$ metal orbital with the A_{1g} ring M.O.; $4p_z$ with the A_{2u} ring M.O.; and $4p_x$ and $4p_y$ with the E_{1u} ring M.O. Interactions between orbitals of symmetry E_{1g} and between those of symmetry E_{2g} must also be considered. The E_{2u} M.O.s on the rings are M.O.s for the whole molecule because there are no metal orbitals of this symmetry with which they can combine.

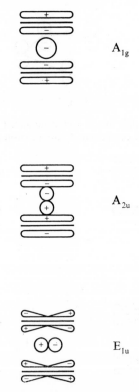

FIG. 20.8. Symmetry-allowed molecular orbitals which are major contributors to the bonding in ferrocene.

Some of the first theoretical treatments of ferrocene[36,37] gave an essentially qualitative account of the bonding based on symmetry arguments. Moffitt concluded that the primary source of bonding between iron and the cyclopentadienyl rings was the interaction between the orbitals, of E_{1g} symmetry. He regarded other interactions, for example, those involving E_{2g} or E_{1u} orbitals, as of only secondary importance so that the bonding would be essentially a single covalent bond between the metal and each cyclopentadienyl ring. Dunitz

198

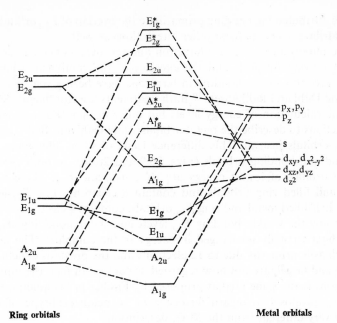

FIG. 20.9. Energy level diagram for ferrocene according to Shustorovich and Dyatkina.[38]

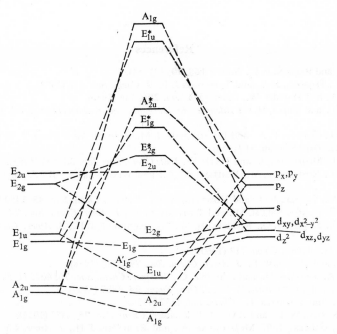

FIG. 20.10. Energy level diagram for ferrocene according to Dahl and Ballhausen.[39]

and Orgel also attributed the bonding primarily to the overlap of E_{1g} orbitals but recognized that some contributions come from other interactions as well.

The relative importance of each interaction allowed by symmetry is determined by the calculation of overlap integrals and the energies of the combined orbitals. Two of the most detailed accounts of the bonding in ferrocene have been given by Shustorovich and Dyatkina[38] and Dahl and Ballhausen.[39] Both treatments use Roothaan's SCF–LCAO–MO method for the calculation of energy levels but differ in that Shustorovich and Dyatkina use Slater functions to describe the metal orbitals while Dahl and Ballhausen use Watson's Hartree–Fock orbitals instead. This difference leads to some differences in the ordering of bonding and anti-bonding orbitals (Figs. 20.9 and 20.10). In both cases, however, the three bonding orbitals of lowest energy are A_{1g}, A_{2u}, and E_{1u}. These are formed from empty metal and filled ring orbitals and contain 8 electrons. The E_{1g} level, formed by overlap of the half-filled metal and ring E_{1g} orbitals, contains 4 electrons and the remaining 6 electrons occupy the weakly bonding E_{2g} levels and the nonbonding A_{1g} level (this last is the $3d_{z^2}$ orbital virtually unchanged in energy in the complex). The bonding between iron and the rings is primarily due to 12 electrons and the previous conclusions of Moffitt and of Dunitz and Orgel, are not now regarded as correct. In fact, the number of strongly bonding electrons is the same as that proposed by Fischer in his qualitative valence-bond description. His proposal to regard ferrocene as an octahedral-type of complex is not fundamentally very different from the M.O. description.

It is interesting to note that the number of electrons required by iron to reach the krypton configuration now becomes, according to M.O. theory, the number required to fill all the strongly and weakly bonding molecular orbitals in the complex.

Detailed treatments of the physical properties and energy levels of ferrocene according to various M.O. theories have been published but are beyond the scope of this work.

References

1. KEALY, T. J., and PAUSON, P. L., *Nature* **168**, 1039 (1951).
2. MILLER, S. A., TEBBOTH, J. A., and TREMAINE, J. F., *J. Chem. Soc.* 632 (1952).
3. FISCHER, E. O., and HAFNER, W., *Z. anorg. Chem.* **286**, 146 (1956).
4. FISCHER, E. O., and FRITZ, H. P., *Advances in Inorganic and Radiochemistry*, vol. I, p. 55, Academic Press, 1959.
5. WILKINSON, G., COTTON, F. A., and BIRMINGHAM, J. M., *J. Inorg. Nucl. Chem.* **2**, 95 (1956).
6. WILKINSON, G., *Org. Syn.* **36**, 31 (1956).
7. WATANABE, H., MOTOYAMA, I., and HATA, K., *Bull. Chem. Soc. Jap.* **38**, 853 (1965).
8. WILKINSON, G., ROSENBLUM, M., WHITING, M. C., and WOODWARD, R. B., *J. Amer. Chem. Soc.* **74**, 2125 (1952).
9. ROSENBLUM, M., SANTER, J. O., and HOWELLS, W. G., *J. Amer. Chem. Soc.* **85**, 1450 (1963).
10. WOODWARD, R. B., ROSENBLUM, M., and WHITING, M. C., *J. Amer. Chem. Soc.* **74**, 3458 (1952).
11. CURPHEY, T. J., SANTER, J. O., ROSENBLUM, M., and RICHARDS, J. H., *J. Amer. Chem. Soc.* **82**, 5249 (1960).
12. GREEN, M. L. H., *Angew. Chem.* 719 (1960).
13. TRIFAN, D. S., and NICKOLAS, L., *J. Amer. Chem. Soc.* **79**, 2746 (1958).
14. BENKESER, R. A., GOGGIN, D., and SCHROLL, G., *J. Amer. Chem. Soc.* **76**, 4025 (1954).
15. ACTON, E. M., and SILVERSTONE, R. M., *J. Org. Chem.* **24**, 1487 (1959).
16. HELLING, J. F., and SHECHTER, H., *Chem. Ind.* 1157 (1959).
17. KAPLAN, L., KESTER, W. L., and KATZ, J. J., *J. Amer. Chem. Soc.* **75**, 6357 (1952).
18. FRAENKEL, G., CARTER, R. E., McLACHLAN, A., and RICHARDS, J. H., *J. Amer. Chem. Soc.* **82**, 5846 (1960).
19. RINEHART, K. L., BUBLITZ, D. E., and GUSTAFSON, D. H., *J. Amer. Chem. Soc.* **85**, 970 (1963).

20. Siebold, E. A., and Sutton, L. E., *J. Chem. Phys.* **23**, 1967 (1955).
21. Bohn, R. K., and Haaland, A., *J. Organomet. Chem.* **5**, 470 (1966).
22. Richmond, H. H., and Freiser, H., *J. Amer. Chem. Soc.* **77**, 2022 (1955).
23. Dunitz, J. D., and Orgel, L. E., *Nature* **171**, 121 (1953).
24. Dunitz, J. D., Orgel, L. E., and Rich, A., *Acta Cryst.* **9**, 373 (1956).
25. Edwards, J. W., Kington, G. L., and Mason, R., *Trans. Faraday Soc.* **56**, 660 (1960).
26. Holm, C. H., and Ibers, J. A., *J. Chem. Phys.* **30**, 885 (1959).
27. Mulay, L. N., and Attalla, A., *J. Amer. Chem. Soc.* **85**, 702 (1963).
28. Lippincott, E. R., and Nelson, R. D., *Spectrochim. Acta* **10**, 307 (1958).
29. Fritz, H. P., *Advances in Organomet. Chem.* **1**, 267 (1964).
30. Lippincott, E. R., and Nelson, R. D., *J. Chem. Phys.* **21**, 1307 (1953).
31. Winter, W. K., Curnutte, B., and Whitcomb, S. E., *Spectrochim. Acta* **15**, 1085 (1959).
32. Cotton, F. A., *Modern Coordination Chemistry* 357 (1960).
33. Long, T. V., and Huege, F. R., *Chem. Comm.* 1239 (1968).
34. Wertheim, G. K., and Herber, R. H., *J. Chem. Phys.* **38**, 2106 (1963).
35. Fischer, E. O., and Pfab, W., *Z. Naturforsch.* 7b, 377 (1952).
36. Moffitt, W., *J. Amer. Chem. Soc.* **76**, 3386 (1954).
37. Dunitz, J. D., and Orgel, L. E., *J. Chem. Phys.* **23**, 954 (1955).
38. Shustorovich, E. M., and Dyatkina, M. E., *Doklady Akad. Nauk. SSSR* **128**, 1234 (1959) and **133**, 141 (1960).
39. Dahl, J. P., and Ballhausen, C. J., *Kgl. Danske Videnskab. Selskab, Mat-fys. Medd.* No. 5, 33 (1961).

CHAPTER 21

BIS(PYRIDINE)COBALT(II) CHLORIDE

Introduction

Many metals form complexes in which nitrogen-containing heterocyclic compounds act as ligands, pyridine being, in this context, probably the most widely studied ligand. For example, an extensive series of complexes of general formula $M^{II}py_2X_2$ (M = transition metal, py = pyridine and X = halogen or pseudohalogen) is formed by most metals of the first transition series.

The pyridine complexes of cobalt(II) halides have a particular feature of interest. Some of them are known to exist in two different forms which differ markedly in colour and other physical properties, such as their magnetic susceptibility. For example, dichloro-bis-(pyridine)cobalt(II), $Copy_2Cl_2$, is known in two solid modifications, α and β. These are respectively violet and blue in colour and their physico-chemical properties have been fully characterized.

Preparation

The violet isomer was first prepared by Reitzenstein[1] in 1894. It is easily made by trituration of anhydrous cobalt chloride (1 mole) with pyridine (2 moles) followed by extraction of the mixture with hot ethanol. When the ethanol extract is cooled, pale violet crystals of α-$Copy_2Cl_2$ separate. The complex can be purified by recrystallization from ethanol.

The other isomer was first prepared in 1927 by Hantzsch.[2] A simple method of preparation is by dissolving the α-form in hot chloroform, then adding petroleum ether whereupon blue crystals of β-$Copy_2Cl_2$ separate out. The blue form reverts to the violet on standing in air for a few hours. The change appears to be autocatalytic and is accelerated by the presence of moisture.

The β-form is also obtained when α-$Copy_2Cl_2$ is heated in a sealed tube at 110–120°C for a few hours.

Both isomers are decomposed by water, forming cobalt(II) chloride and pyridine. They dissolve in acetone, ethanol and chloroform to give deep-blue solutions. Evaporation of the solvent at or near room temperature gives the α-isomer irrespective of which isomer is originally dissolved. Evaporation at higher temperatures gives the β-isomer.

From ebullioscopic measurements, it has been concluded that the monomeric molecule, $Copy_2Cl_2$, exists in nitrobenzene, chloroform and bromoform solutions. When either isomer is dissolved in a particular solvent, the same absorption spectrum is observed for both solutions. It appears, from this evidence, that the violet form does not exist in solution and that it is probably a solid state polymer which breaks down on dissolution.

The transition $\alpha \rightarrow \beta$ has been studied by several thermal methods,[3,4] namely differential thermal analysis, dynamic reflectance spectroscopy (the measurement of changes in reflectance at a fixed wavelength as the temperature of a solid is increased) and dilatometry (the measurement of the volume change accompanying the transition). This work has confirmed that the transition occurs between 110° and 125°C. From differential scanning calorimetry, the heat of transition has been found to be $13\cdot4 \pm 0\cdot4$ kJ mol^{-1}.

Beyond 125°C, β-Copy$_2$Cl$_2$ decomposes[5] in a series of well-defined steps:

$$\text{Copy}_2\text{Cl}_2(s) \xrightarrow{210\,°C} \text{CopyCl}_2(l) + \text{py}(g)$$

$$\text{CopyCl}_2(l) \xrightarrow{250\,°C} \text{Copy}_{2/3}\text{Cl}_2(l) + \tfrac{1}{3}\text{py}(g)$$

$$\text{Copy}_{2/3}\text{Cl}_2(l) \xrightarrow{350\,°C} \text{CoCl}_2(s) + \tfrac{2}{3}\text{py}(g)$$

The intermediate products have been characterized by diffuse reflectance spectroscopy. The spectra of CopyCl$_2$ and Copy$_{2/3}$Cl$_2$ show bands between 500 and 650 nm and another one near 800 nm. The latter is found also in the spectrum of solid cobalt(II) chloride and is regarded as typical of octahedrally coordinated cobalt. Although definitive structural analyses of the intermediate products of decomposition have not yet been reported, it is very probable they contain octahedral cobalt. This coordination could be derived from the polymeric chain structure of α-Copy$_2$Cl$_2$ by removal of some pyridine molecules, leaving chains which then join up together by way of chlorine atoms acting as bridges between three metal atoms.

The heat of dissociation[6] of β-Copy$_2$Cl$_2$, according to the equation: β-Copy$_2$Cl$_2$(s) $\rightarrow$ CoCl$_2$(s) + 2py(g) is $119\cdot7 \pm 2\cdot1$ kJ mol^{-1}. From calorimetric studies on solutions,[7] the heat of dissociation of α-Copy$_2$Cl$_2$ according to the equation:

$$\alpha\text{-Copy}_2\text{Cl}_2(s) \rightarrow \text{CoCl}_2(s) + 2\text{py}(g) \text{ is } 189\cdot5 \pm 2\cdot1 \text{ kJ mol}^{-1}.$$

Magnetic susceptibility data at 20°C on the two solid forms[8] and on solutions[9] in various solvents have been reported. For solid α-Copy$_2$Cl$_2$, $\mu_{\text{eff}} = 5\cdot15$ B.M. For solid β-Copy$_2$Cl$_2$, $\mu_{\text{eff}} = 4\cdot42$ B.M. In nitrobenzene solution, $\mu_{\text{eff}} = 4\cdot52$ B.M. These data have been related to the probable stereochemistry of cobalt in the two solid isomers. In general, the magnetic moments of tetrahedral cobalt(II) complexes are in the range 4·4 to 4·8 B.M. whilst those of octahedral complexes are between 4·8 and 5·2 B.M. On the assumption of spin-only contributions to the magnetic moment, the calculated value of μ_{eff} for spin-free cobalt(II) complexes is 3·88 B.M. (since Co^{II} is a d^7 ion) whether the complex is tetrahedral or octahedral. The experimental values are higher for both stereochemistries due to contributions to the paramagnetism from the orbital angular momentum of the unpaired electrons. These data indicate that α-Copy$_2$Cl$_2$ is an octahedral complex, presumably polymeric, and that β-Copy$_2$Cl$_2$ is a tetrahedral monomer.[9,10]

Other cobalt(II) complexes have been examined and their properties related to those of the isomers of Copy$_2$Cl$_2$. For example, only the blue forms of Copy$_2$Br$_2$ and Copy$_2$I$_2$ appear to exist in the solid state. Their magnetic moments at 20°C are respectively 4·50 and 4·47 B.M. and these indicate a tetrahedral configuration around cobalt. The thiocyanate, Copy$_2$(SCN)$_2$, exists in the solid as a violet form which gives blue solutions. For the solid, $\mu_{\text{eff}} = 5\cdot10$ B.M., and for its solution in nitrobenzene, $\mu_{\text{eff}} = 4\cdot50$ B.M. Evidently, this shows the same kind of isomerism as $Copy_2Cl_2$ although only the octahedral form of the thiocyanate is known in the solid state.

Structure

The α-isomer is monoclinic and is composed of polymeric chains[11] which run parallel to the needle axis of the crystal. The coordination about each cobalt is octahedral, the metal being surrounded by four chlorines (Co,Cl = 0·249 nm) and two nitrogens (Co,N = 0·214 nm). This structure is illustrated in Fig. 21.1. At least three other dichloro-bis(pyridine)

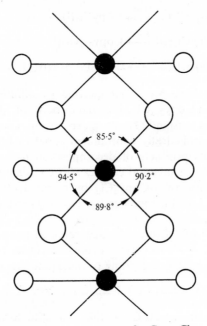

FIG. 21.1. The structure of α-$Copy_2Cl_2$.

metal(II) compounds (those of manganese, copper and mercury) are known to have a similar polymeric structure and it appears to represent a particularly stable structure for this kind of complex. The copper(II) complex is less regular in its structure for each metal has two nearer chlorines (at a distance of 0·228 nm) and two further away (at 0·305 nm).

The β-isomer is believed to contain discrete tetrahedral molecules. Its structure does not appear to have been determined directly but this conclusion is based on resemblances to other compounds known to contain tetrahedral cobalt(II). For example, X-ray analysis[12] of the blue compound, Co(p-toluidine) dichloride, shows this to be tetrahedral.

Gill and Nyholm[9] have compared three halides of cobalt in relation to their ability to form octahedral complexes with pyridine. Thus $CoCl_2$ forms both octahedral and tetrahedral isomers, the former being the more stable in the solid state. In contrast, only the blue tetrahedral forms of $Copy_2Br_2$ and $Copy_2I_2$ have been isolated. Of these, the bromide can absorb two water molecules to form a hydrate, the violet colour of which suggests octahedral coordination. The magnetic moment of solid $Copy_2Br_2 2H_2O$ is 5·0 B.M., but, on solution in nitrobenzene, this falls to 4·52 B.M., a value corresponding to a tetrahedral complex. The iodide appears to show no tendency to increase the number of ligands bound to cobalt beyond four.

The coordination number of cobalt(II) may be considered in relation to the polarizability of the halogen ligand. In forming its complexes, the metal ion accepts electronic charge from the ligands to such an extent that, in accordance with the Electroneutrality Principle enunciated by Pauling, its charge is reduced to almost zero. For a ligand like chloride ion, which is not easily polarized, relatively little charge is transferred to the metal compared with the state of affairs when a highly polarizable ligand like iodide is involved. Hence the preference for the higher coordination number with chloride and the increasing tendency for a lower number as we pass from bromide to iodide.

Spectroscopic Studies

In recent years, the technique of far infrared spectroscopy has been applied to an increasing extent in the study of the structures of metal complexes. The spectral region concerned ($600–150$ cm^{-1}) is where absorption bands due to the vibrations of metal–ligand bonds usually occur, the number of such bands being dependent on the stereochemistry of the complex. One of the earliest successes[13,14] of far infrared spectroscopy was, in fact, the clear-cut differentiation between the octahedral and tetrahedral isomers of Copy$_2$Cl$_2$. The β-isomer gives two absorption bands, at 344 and 304 cm^{-1}, which correspond respectively with symmetric and asymmetric stretching frequencies of the metal–halogen bonds. The polymeric α-isomer shows, in contrast, only one broad band, at around 240 cm^{-1}. Other absorption bands are, of course, observed in this spectral region due to ring vibrations of pyridine and the metal–nitrogen stretching mode but these are readily identified because they show little variation in frequency in the series of compounds, Copy$_2$X$_2$, in which the nature of X, the halogen, is varied.

The electronic absorption spectra of Copy$_2$Cl$_2$ and related complexes have been measured[15] in solution and in the crystalline state. In solvents such as isopropanol, acetone, chloroform or bromoform, β-Copy$_2$Cl$_2$ shows one complex absorption band, composed of four overlapping maxima, in the region $15,000–17,000$ cm^{-1} and a second band in the near infrared region between 6000 and 8000 cm^{-1}. There is further absorption at frequencies above 20,000 cm^{-1}. These features are shown in Fig. 21.2, which also illustrates that β-Copy$_2$Cl$_2$ in solution does not obey Beer's Law. Marked changes with concentration in the frequencies of maximum absorption are observed, probably because of molecular association.

The ground state of Co(II) (a d^7 ion) in a tetrahedral field is $^4A_2(F)$ and spin-allowed transitions can occur from this to the excited states $^4T_2(F)$, $^4T_1(F)$ and $^4T_1(P)$. (The atomic state from which the level is derived is added in brackets after the group theoretical designation.) The order of energy of these states is given in Fig. 21.3, a simplified energy level diagram for tetrahedral Co(II). The transition of lowest energy ν_1, $^4T_2(F) \leftarrow {}^4A_2(F)$, should occur in the region $3000–5000$ cm^{-1} but appears to have been rarely observed. The transition ν_3, $^4T_1(P) \leftarrow {}^4A_2(F)$, is responsible for the absorption between 15,000 and 20,000 cm^{-1} which results in the blue colour frequently characteristic of tetrahedral cobalt(II) complexes. Intermediate in energy between these transitions is ν_2, $^4T_1(F) \leftarrow {}^4A_2(F)$, leading to absorption bands in the near infrared.

The absorption bands observed in solutions of β-Copy$_2$Cl$_2$ correspond broadly with transitions ν_2 and ν_3. However, it must be emphasized that such a complex does not have

205

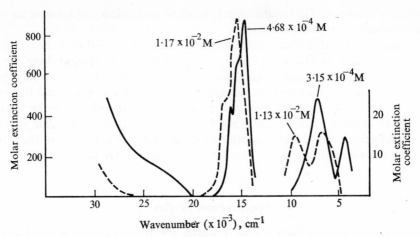

FIG. 21.2. The absorption spectrum of β-Copy$_2$Cl$_2$ in bromoform solutions of different concentrations (reproduced with permission from Ferguson, J., *J. Chem. Phys.* **32**, 528 (1960)).

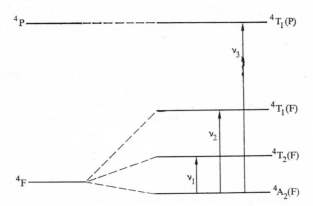

FIG. 21.3. Simplified energy level diagram for tetrahedral cobalt(II).

true tetrahedral symmetry because the presence of more than one kind of ligand in the coordination sphere of cobalt reduces its symmetry at least to C_{2v}. Further splitting of the energy levels for a d^7 ion occurs in this lower symmetry and affects the structure and energy of each absorption band. The above description of the absorption spectrum of Copy$_2$Cl$_2$ is therefore very much a simplified account of the real situation. This is illustrated by the absorption bands observed for β-Copy$_2$Cl$_2$ in the near infrared region. Although Fig. 21.3 suggests only one transition (ν_2), three maxima have been observed[16] in the diffuse reflectance spectrum of this compound. These are located respectively at 9430, 7407 and 6250 cm^{-1}. The detailed analysis of these bands is a complicated matter, but it is worth noting that their observation serves to differentiate a tetrahedral from an octahedral complex because the latter shows only one absorption band in the same region. For example, the reflectance spectrum of α-Copy$_2$Cl$_2$ shows one band only, at 9260 cm^{-1}. Indeed, absorption spectroscopy in this spectral region is potentially of great utility[16] for it is possible to observe transitions due to ligand-field effects without interference from charge-transfer absorption which often obscures bands in the visible or ultraviolet spectrum.

References

1. REITZENSTEIN, F., *Ann.* **282**, 273 (1894).
2. HANTZSCH, A., *Z. anorg. Chem.* **159**, 273 (1927).
3. WENDLANDT, W. W., *Chem. Analyst* **53**, 71 (1964).
4. WENDLANDT, W. W., *Analyt. Chim. Acta* **33**, 98 (1965).
5. ALLEN, J. R., BROWN, D. H., NUTTALL, R. H., and SHARP, D. W. A., *J. Inorg. Nucl. Chem.* **26**, 1895 (1964).
6. BEECH, G., MORTIMER, C. T., and TYLER, E. G., *J. Chem. Soc. A* 925 (1967).
7. BEECH, G., ASHCROFT, S. J., and MORTIMER, C. T., *J. Chem. Soc. A* 929 (1967).
8. BARKWORTH, E. D. P., and SUGDEN, S., *Nature* **139**, 374 (1937).
9. GILL, N. S., NYHOLM, R. S., BARCLAY, G. A., CHRISTIE, T. I., and PAULING, P. J., *J. Inorg. Nucl. Chem.* **18**, 88 (1961).
10. MELLOR, D. P., and CORYELL, C. D., *J. Amer. Chem. Soc.* **60**, 1786 (1938).
11. DUNITZ, J., *Acta Cryst.* **10**, 307 (1957).
12. BOKAI, G. B., MALINOWSKI, T. I., and ABLOV, A. V., *Kristallografiya* **1**, 49 (1956).
13. CLARK, R. J. H., and WILLIAMS, C. S., *Chem. Ind.* 1317 (1964).
14. CLARK, R. J. H., and WILLIAMS, C. S., *Inorg. Chem.* **4**, 350 (1965).
15. FERGUSON, J., *J. Chem. Phys.* **32**, 528 (1960).
16. GRADDON, D. P., and MOCKLER, G. M., *Aust. J. Chem.* **21**, 1775 (1968).

CHAPTER 22

TRIS(ETHYLENEDIAMINE)COBALT(III) COMPLEXES

Introduction

The concept of molecules as three-dimensional bodies was first introduced in the field of organic chemistry and the recognition that, in compounds of four-covalent carbon, its valences are tetrahedrally disposed, formed the starting-point for the development of organic stereochemistry. The suggestion was first made by van't Hoff that the three-dimensional shape of organic molecules provides an insight into the phenomenon of optical isomerism and he pointed out that many of the compounds which show this possess an asymmetric carbon atom. This is an important but not the sole cause of optical activity among organic compounds. This property should be associated with any molecule which is dissymmetric by virtue of forming a mirror-image which is not superimposable. A dissymmetric structure has no plane or centre of symmetry but does possess certain symmetry axes.

The arguments used to support the tetrahedral disposition of carbon valences were subsequently used by Alfred Werner to establish the stereochemistry of metal complexes of the type MX_4 and MX_6. Werner proposed that in the case of a complex MX_6 "if we think of the metal atom as the centre of the system, we can most simply place the molecules bound to it at the corners of an octahedron". Much of the experimental evidence on which Werner's theory is based was derived from studies on the complexes of cobalt(III) in which the cobalt atom shows a characteristic coordination number of six.

Clearly the group MX_6 has very high symmetry but this is progressively reduced by the successive replacement of X by different ligands. The essential condition for optical isomerism in coordination compounds is the same as for organic substances, i.e. the complex should lack a plane or centre of symmetry. This condition is observed, for example, in the complex MX_3ABC when the three dissimilar ligands A, B and C, occupy the three corners of a triangular face of the octahedron. The number of different ligands needed to produce dissymmetry in complexes is greatly reduced if these are of the chelating type. Ethylenediamine ("en"), $NH_2CH_2CH_2NH_2$, was one of the first chelating ligands to be studied. The nitrogen atoms of one molecule can occupy two coordination positions and, in so doing, form a five-membered heterocyclic ring containing the metal atom. Werner recognized the ability of ethylenediamine to behave in this manner and attributed the optical isomerism shown by complexes of the type $[Coen_2X_2]X$ to the existence for this of two non-superimposable configurations (enantiomers) which are the direct consequence of an octahedral disposition of the six ligand atoms around the cobalt.

The final experimental confirmation of Werner's theory came with the resolution of complexes such as tris(ethylenediamine)cobalt(III) salts into their optical isomers. The optical activity here arises from the two configurations I and II (Fig. 22.1) which are

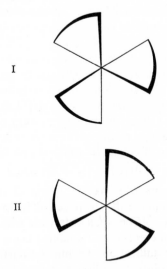

FIG. 22.1. The enantiomers of $Coen_3^{3+}$.

related as object and mirror-image. The ion is dissymmetric and possesses a three-fold rotation axis of symmetry (C_3) and three two-fold rotation axes (C_2) perpendicular to this. The distinction between the two configurations is clear if reference is made to the nature of the rotations possible about these axes. In I the C_3 axis passing through the metal is perpendicular to the plane of the paper and a rotation about this in a clockwise direction causes the structure to behave in the same way as a left-handed propeller or screw by showing an apparent movement towards the observer. This movement can be designated as sinistral with respect to the C_3 axis (abbreviated to $S(C_3)$). If I is rotated about one of its C_2 axes in a clockwise direction, it moves as a right-handed propeller or screw and is designated as $R(C_2)$. I is therefore completely described by $S(C_3)R(C_2)$ and correspondingly the enantiomer II by $R(C_3)S(C_2)$.

The enantiomers of $Coen_3^{3+}$ rotate the plane of polarized light in opposite directions. That which is dextrorotatory at the Na_D line is designated by the prefix $(+)$ and that which is laevorotatory by $(-)$. For many years it was not possible to relate these designations unambiguously with the configurations I and II. Recently, the absolute configuration of I has been determined by the technique of anomalous diffraction of X-rays. This result has far-reaching implications for it makes possible the determination of the configurations of many other optically-active complexes. In order to do this, the investigation of the phenomena of optical rotatory dispersion and circular dichroism in metal chelates has assumed special importance and much recently-published work is devoted to the interpretation of these effects in terms of the electronic structure and configuration of the species studied.

Preparation

Many cobalt(III) ammines are converted by reaction with aqueous ethylenediamine to tris(ethylenediamine)cobalt(III) salts. Thus Jörgensen prepared the chloride by heating $[Co(NH_3)_5Cl]Cl_2$ with aqueous ethylenediamine.[1] More conveniently, this compound is made by the method originally used by Grossman and Schück[2] in which a mixture of cobalt(II) chloride, ethylenediamine and water is oxidized. The salt can be isolated from the reaction mixture by evaporation of some of the solvent whereupon orange-yellow crystals begin to separate. The bulk of the product can be precipitated by the addition of concentrated hydrochloric acid and ethanol.

When the separate enantiomers are required there is no need to isolate the unresolved complex as a solid and the separation can be effected by applying the classical method of diastereoisomer formation to the solution. In the recommended method[3] the complex is made by oxidation of an aqueous solution of cobalt(II) sulphate (1 mol), ethylenediamine (3 mol) and hydrochloric acid (1 mol) by passing a rapid current of air through it for 3 to 4 hr. Activated charcoal is added to the reaction mixture to ensure the rapid oxidation of the ethylenediamine complex of cobalt(II) which is first formed.

$$4CoSO_4 + 12en + 4HCl + O_2 \rightarrow 4[Coen_3]ClSO_4 + 2H_2O$$

The resolution was first performed by Werner[4] using silver (+)tartrate as the resolving agent. This converts the racemic salt $(\pm)[Coen_3]Cl_3$ to a pair of diastereoisomers, $(+)[Coen_3]Cl(+)$tartrate and $(-)[Coen_3]Cl(+)$tartrate which are not longer enantiomorphous. Their solubilities are different, the first-named being the less soluble, and they can be separated by fractional crystallization. Barium (+)tartrate is preferable as resolving agent in view of its greater stability and lower cost than the silver salt.

After the addition of barium (+)tartrate to the solution containing $[Coen_3]^{3+}$, the barium sulphate formed is removed by filtration and the filtrate concentrated until $(+)[Coen_3]Cl(+)$tartrate crystallizes out. This can be removed and the resolving agent removed by dissolving the complex in water and treating the solution with sodium iodide to precipitate out $(+)[Coen_3]I_3$. The mother liquor remaining after removal of the less soluble diastereisomer contains $(-)[Coen_3]^{3+}$ and some racemate. Both are precipitated by the addition of sodium iodide. The impure iodide is extracted with water leaving the racemate iodide undissolved and pure $(-)[Coen_3]I_3$ is precipitated by the addition of sodium iodide to the aqueous extract.

The enantiomers $(+)$ and $(-)[Coen_3]I_3$ are quite stable in the solid state and in solution. When heated to 110°C in a sealed tube, the solid racemizes. Racemization is a slow process in solution, even at 100°C, unless catalysts such as charcoal or finely-divided metals are present.

Alternatively, the optical isomers of $[Coen_3]^{3+}$ can be made by the method of partial asymmetric synthesis.[5] This involves carrying out the reactions to form the complex in the presence of a resolving agent, a procedure which generally gives high yields of one enantiomer at the expense of the other. A mixture of cobalt(II) (+)tartrate (1 mol), ethylenediamine (3 mol) and hydrochloric acid (1 mol) in aqueous ethanol is aerially oxidized in the presence of activated charcoal. In this system, some $[Coen_3]^{2+}$ is present until oxidation is complete and this ion causes racemization, by an electron-transfer process, of the $(-)[Coen_3]^{3+}$ ion in solution. Thus the more soluble enantiomer racemizes and precipitation of the less soluble diastereoisomer, $(+)[Coen_3]Cl(+)$tartrate, is promoted. As in the previous method, the resolving agent can be removed by conversion of the complex to the iodide. Yields of up to 65% of $(+)[Coen_3]I_3$ are obtained in this way.

Historically, organic resolving agents were used for dissymmetric inorganic complexes. In addition to (+)tartaric acid, (+)camphor- and bromocamphor-sulphonic acids were used in the resolution of cations and bases like (−)strychnine, (−)brucine and (−)cinchonine for anions. More recently, it has been demonstrated that resolved metal complexes can themselves act as resolving agents. Thus Dwyer and Sargeson[6] showed that $Ni(phen)_3^{2+}$ (phen = 1:10 phenanthroline) can be used to resolve the tris(oxalato) complexes of Co(III), Cr(III) and Rh(III). This procedure has the great advantage that it avoids the difficulties of working with extremely toxic materials like strychnine. It has now been found by Vaughn and co-workers[7] that $(−)[Coen_3]^{3+}$ is also effective for the resolution of these three oxalato complexes.

The Stereochemistry of tris(ethylenediamine)cobalt(III) Complexes

To reach an understanding of the stereochemistry of metal chelates, there are two distinct but related problems to be considered. These, which have been discussed in detail by Corey and Bailar,[8] are:

(1) The orientation of donor atoms about the central metal ion, and
(2) The spatial arrangements which can be assumed by the individual chelate rings and their relative stabilities.

The configuration of donor atoms around the metal ion is not necessarily the same as in the corresponding non-chelate complexes. Some distortion from the regular orientation will result in most cases because of the geometrical requirements of the chelate ring.

The first proposal that a five-membered ring formed by chelated ethylenediamine is not necessarily planar was made in 1933 with reference to the bis(ethylenediamine)platinum(II) cation.[9] Kobayashi, from a study of the optical rotatory dispersion of $(+)[Coen_3]Br_3$, concluded[10] that in this compound as well the coordinated ethylenediamine rings are puckered. The subsequent X-ray diffraction study[11] of $[Coen_3]Cl_3 \cdot 3H_2O$ confirmed that the chelate rings are not planar and showed that the ligand coordinates in a *gauche* configuration. The dimensions of the chelate ring reported in this study are:

$$C-C = 0.154 \text{ nm} \qquad \widehat{CoNC} = 109.5°$$
$$C-N = 0.147 \text{ nm} \qquad \widehat{NCC} = 109.0°$$
$$Co-N = 0.200 \text{ nm} \qquad \widehat{NCoN} = 87.4°$$
$$\theta = 48°$$

The $\widehat{NCoN}$ angle shows that there is a slight distortion from a regular octahedral arrangement of nitrogens around the cobalt. The angle θ is the azimuthal angle between the projections of the two C,N bonds viewed down the C,C axis. It is a measure of the extent of ring non-planarity ($\theta = 0$ for a planar ring).

Chelated ethylenediamine is a ring system comparable with cyclopentane and the techniques used for the stereochemical analysis of carbocyclic systems are also applicable to chelate rings. Corey and Bailar[8] used the methods of conformational analysis in their examination of the ring systems in $[Coen_3]^{3+}$ and found that the bond angles and lengths calculated by vector analysis agreed almost exactly with those found experimentally by X-ray diffraction. An important feature of the conformation of the chelate ring is that there is virtually complete staggering of the hydrogens on adjacent ring atoms, an arrangement which minimizes the energy of the system.

The non-planarity of the chelate rings leads to further isomerism in complexes like $[Coen_3]^{3+}$. Ethylenediamine in a *gauche* form exists as two rotational isomers which are enantiomorphous with each other (Fig. 22.2). These are designated δ and λ in accordance with IUPAC recommendation (corresponding respectively with the k and k' notations of Corey and Bailar[8]). Ethylenediamine itself cannot be resolved into these enantiomers because the activation energy for isomerization is too small but the ligand can adopt either conformation in chelation (Fig. 22.3). For one configuration of an octahedral tris(ethylene-diamine) complex, there are four possible arrangements of the three rings. These are

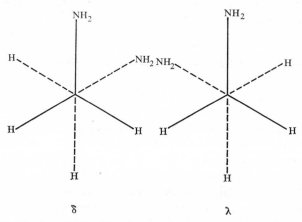

FIG. 22.2. The δ- and λ-configurations of ethylenediamine in a *gauche* form.

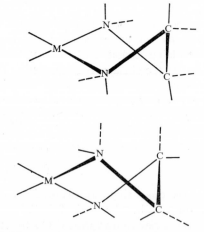

FIG. 22.3. The conformations of coordinated ethylenediamine.

$\delta\delta\delta$, $\delta\delta\lambda$, $\delta\lambda\lambda$ and $\lambda\lambda\lambda$. These structures are expected to have different free energies and stabilities. Corey and Bailar considered the non-bonded interactions between the following pairs of atoms: (1) H,H; (2) H,C (involving the two carbon atoms of a ring and the axial hydrogen of the NH_2 groups) and (3) hydrogens of the donor NH_2 groups. They concluded that for the configuration represented by I (and designated Δ in accordance with IUPAC recommendation) the sequence of stability is:

$$\Delta(\delta\delta\delta) > \Delta(\delta\delta\lambda) > \Delta(\delta\lambda\lambda) > \Delta(\lambda\lambda\lambda)$$

The energy difference between the two extremes, $\Delta(\delta\delta\delta)$ and $\Delta(\lambda\lambda\lambda)$, was estimated as 7.5 kJ mol^{-1} (2.5 kJ mol^{-1} for each ligand) largely because of less severe H,C interactions in the former isomer. Although these four isomers cannot be resolved because of the small energy differences between them, it has been shown[12] from X-ray diffraction analysis that the most stable form of Δ-Coen$_3$$^{3+}$ is $\Delta(\delta\delta\delta)$ in the solid state.

The difference between $\Delta(\delta\delta\delta)$ and $\Delta(\lambda\lambda\lambda)$ in terms of their stereochemistries is that, in the first compound, the direction of the central C–C bond in each ethylenediamine molecule is approximately parallel to the three-fold axis of the complex whereas in the second, the C–C bonds are slanted obliquely relative to this. This difference is reflected in the nomenclature 'lel' and 'ob' respectively given to these isomers by Corey and Bailar. The two isomers presumably have different stabilities because the repulsions between the ligands are different.

For the configuration II (designated Λ) there are again four possible configurations:

$$\Lambda(\lambda\lambda\lambda),\ \Lambda(\lambda\lambda\delta),\ \Lambda(\lambda\delta\delta)\ \text{and}\ \Lambda(\delta\delta\delta)$$

and the order of stabilities falls from $\Lambda(\delta\delta\delta)$ to $\Lambda(\lambda\lambda\lambda)$.

The energy difference between the most and least stable isomers has also been determined experimentally.[13,14] The method involved the equilibration of a mixture of cobalt(III), ethylenediamine and ($-$)propylenediamine and the determination of the ratios of the Δ and Λ isomers of the mixed complexes [Coen$_2$($-$)pn]$^{3+}$ and [Coen(($-$)pn)$_2$]$^{3+}$. On the assumptions that, for the Δ configuration, the ethylenediamine chelate ring adopts the δ conformation and, for the Λ, it adopts the λ conformation, and that the ($-$)propylenediamine chelate ring retains its λ conformation irrespective of the configuration, the relative energies found for $\Delta(\delta\delta\delta)$, $\Delta(\delta\delta\lambda)$, $\Delta(\delta\lambda\lambda)$ and $\Delta(\lambda\lambda\lambda)$ were respectively 0, 2.7, 3.3, and 6.7 kJ mol^{-1}.

The basic assumption that the conformation of ethylenediamine is completely fixed by the configuration is regarded by Gollogly and Hawkins[15] as being too restrictive and they propose that it can adopt either conformation. Their re-interpretation of the experimental data concerning energy differences between the various configurations of Coen$_3$$^{3+}$ leads to the conclusion that $\Delta(\delta\delta\delta)$ and $\Delta(\delta\delta\lambda)$, and their counterparts $\Lambda(\lambda\lambda\lambda)$ and $\Lambda(\lambda\lambda\delta)$, are very close in energy and are to be considerably preferred over the other configurations. This means that, in practice, each 'enantiomer' probably consists of a mixture of at least two isomers.

The Absolute Configuration of [Coen₃]³⁺

In general, although the arrangement of atoms in a crystal can be determined by X-ray diffraction, it appeared for many years to be impossible to distinguish between two optical isomers by this technique. In the particular case of [Coen$_3$]$^{3+}$, it was not possible to decide which of the configurations I and II corresponded with ($+$)[Coen$_3$]$^{3+}$. The reason for this is that the phase difference between X-rays diffracted by any pair of atoms is normally independent of the nature of these atoms and depends only on the differences in path length of the rays diffracted by them. Thus if the positions of the two atoms are interchanged, the difference in path length between X-rays diffracted by them remains unchanged and so does the contribution to the intensity of the diffracted rays. This is expressed in Friedel's Law which states the equality of the intensities of the hkl and $\bar{h}\bar{k}\bar{l}$ reflections, i.e. $I(hkl) = I(\bar{h}\bar{k}\bar{l})$. The argument also applies to X-ray diffraction by an asymmetric molecule and its inverted (enantiomorphous) form. The contribution of each to the intensity of diffracted rays is the same and they cannot be distinguished from each other. For $(+)$ and $(-)$ isomers, $I_{(+)}(hkl) = I_{(-)}(hkl)$.

The key to the determination of absolute configuration lies in so arranging the experimental conditions that Friedel's Law is not obeyed. Normally, X-rays interact with the orbital electrons in atoms, are scattered and diffraction is observed. If, however, the energy of the incident radiation is within the range where the inner (K) shell electrons of one atom absorb and undergo orbital transitions, the scattering by this atom is accompanied by a phase change. Friedel's Law is no longer obeyed and the X-rays are scattered anomalously.

The inequality of $I(hkl)$ and $I(\bar{h}\bar{k}\bar{l})$ was first demonstrated experimentally on zinc blende crystals[16],[17] and the first determination of the absolute configuration of an organic molecule (sodium rubidium tartrate dihydrate) by the technique of anomalous diffraction of X-rays was performed in 1951 by J. M. Bijvoet and co-workers.[18] The application of this technique to inorganic complexes led to the establishment[19] of the absolute configuration of $(+)[\text{Coen}_3]^{3+}$. As the wavelength of $\text{CuK}\alpha$ radiation $(\lambda = 0\cdot1542 \text{ nm})$ is a little shorter than the K absorption edge of the cobalt atom $(0\cdot1608 \text{ nm})$ it is suitable for anomalous diffraction. The complex ion I represents the absolute configuration of $(+)[\text{Coen}_3]^{3+}$. More recently, the absolute configuration of $(-)[\text{Co}(-\text{pn})_3]^{3+}$ has also been determined[20] and these two complexes provide reference substances for the deduction of the configurations of many other complexes.

N.M.R. Spectroscopy

The use of high resolution N.M.R. spectroscopy to study the bonding and structure of metal chelates is a relatively recent development. In some instances, it has proved to be an elegant technique for distinguishing between isomers. For example, the tris complexes of benzoylacetone and Co(III) and Rh(III) exist as *cis* and *trans* isomers and they have characteristically different N.M.R. spectra.[21] The application of this technique to $[\text{Coen}_3]^{3+}$ and analogous complexes has thrown some further light on the interrelationships of the various conformational isomers.

The most difficult problem in applying N.M.R. spectroscopy to the study of cobalt(III) complexes has been to find a suitable solvent. Most of them are not soluble in ordinary organic solvents and aqueous solutions present difficulties because of strong resonance due to water and the probability of rapid proton exchange between ligands and water which makes it difficult to pick up signals of the ligand separately. For these reasons, most of the work on $[\text{Coen}_3]^{3+}$ has been carried out with solutions in D_2O, acidified D_2O or in liquid solvents like trifluoroacetic acid and iso-butyric acid.

In D_2O/D_2SO_4 solutions of $[\text{Coen}_3]^{3+}$ salts, the N.M.R. spectrum at 100 MHz shows[22] a broad single peak at higher field and a doublet, largely unresolved, at lower field. As an aid to the assignment of these peaks, the spectrum of the deuterated complex (containing ND_2 instead of NH_2 groups) has also been measured. The signal at lower field disappears and must be due to NH_2 protons in the non-deuterated complex, and the single peak at higher field is therefore assigned to the CH_2 protons. The spectrum resembles that of protonated ethylenediamine which shows lower field resonance due to NH_3^+ and a peak at higher field due to CH_2.

Deuteration of the NH_2 groups in $[\text{Coen}_3]^{3+}$ causes a slight but significant decrease in the half-width of the peak due to CH_2 protons and this suggests that its broadness in the undeuterated complex is due, in part, to spin–spin coupling between the CH_2 and NH_2 protons. Other possible causes of broadening are coupling between the protons and the nuclear spin (7/2) of cobalt and quadrupole relaxation of the ^{14}N nucleus. Yoneda and Morimoto[23] have suggested however that the main cause of band broadening is inherent in the nature of the complex ion. In this, many protons which are magnetically different, are crowded together in a relatively rigid framework around the central metal ion. The situation is thus somewhat like that in a solid, a complete time-averaging of local fields is not realized and a broad band results.

The observation of a single peak only due to the CH_2 protons is regarded[22] as evidence for a rapid conformational exchange whereby they are all rendered magnetically equivalent. Alternatively, if the chelated ethylenediamine molecule is in a rigidly-fixed *gauche* conformation, the failure to resolve the expected fine structure of its N.M.R. spectrum could be the consequence of the operation of one or more of the above broadening effects.

The broad doublet due to NH_2 protons in $[Coen_3]^{3+}$ is found in D_2O/D_2SO_4 and in trifluoroacetic acid solutions of its salts. There are evidently two kinds of magnetically nonequivalent protons present. These can be distinguished[23] in the stereochemistry of the complex ion. Figure 22.4 represents the upper half of the ion viewed along its three-fold

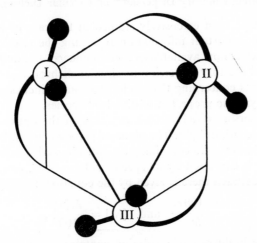

FIG. 22.4. The structure of $[Coen_3]^{3+}$ showing two kinds of protons (small black circles) which are distinguishable by N.M.R. spectroscopy.

axis. On the assumption that the nitrogen has a regular tetrahedral configuration, the six N–H bonds are of two kinds. Three are directed upwards from the plane defined by N(I), N(II) and N(III) and the other three outwards nearly perpendicular to the three-fold axis. The upper three hydrogens form a regular triangle and the lower three a similar larger triangle. The same situation obtains in the lower half of the complex.

In some acidified aqueous solutions, the NH_2 signal appears as a broad single band showing no fine structure. This is believed to be due to a rapid exchange of chelated ethylenediamine from the λ to the δ conformation (and vice versa) with a consequent magnetic equivalence of all NH_2 protons. Such exchange is presumably facilitated by proton dissociation from the NH_2 group which is more likely to occur in aqueous solution than in trifluoroacetic acid.

Two kinds of CH_2 protons exist in the structure represented by Fig. 22.4 and so, theoretically, a doublet should be observed for the CH_2 proton resonance. Failure to observe this could be because the chemical shift is too small or that the CH_2 protons all become equivalent by some movement of the CH_2 group.

Some evidence for the movement of coordinated ethylenediamine has been gathered from observations of the N.M.R. spectrum of $[Coen_3]^{3+}$ in trifluoroacetic acid and iso-butyric acid at different temperatures.[24] In iso-butyric acid, the broad NH_2 doublet found at lower temperatures coalesces above 80°C. This behaviour is consistent with an increasing mobility of the NH_2 groups as the temperature is increased.

The conformational analysis of the $(-)$propylenediamine chelate rings in $[Co((-)pn)_3]^{3+}$ has shown[25] that these have a fixed λ-*gauche* form in aqueous solutions at room temperature. The conformer with this stereochemistry has the methyl group of the propylenediamine molecule directed equatorially with respect to the chelate ring, the alternative arrangement with the methyl group axially directed being less stable. The propylenediamine ring appears to be much less labile to the kind of conformational change which is believed to occur rapidly in the ethylenediamine chelate.

Optical Rotatory Dispersion and Circular Dichroism

These phenomena have been extensively studied in the case of $[Coen_3]^{3+}$ and similar tris chelates and much important information regarding the electronic states and structures of these complexes is now available as a consequence. In addition, the experimental measurement of optical rotatory dispersion or circular dichroism serves as a useful means of investigating problems concerning the mechanisms and kinetics of reaction and of determining the thermodynamic properties of dissolved species.

Optical rotatory dispersion (ORD) is the variation of the optical rotatory power of a medium with the wavelength of light. The plane of plane-polarized light is rotated when it passes through a solution or solid containing an optically-active substance and the optical rotatory power of this is expressed by its specific rotation, $[\alpha]$, or its molecular rotation, $[M]$. The relationship between these is given by the equations:

$$ = \frac{100\,\alpha}{lc} \quad \text{and} \quad [M] = \frac{M[\alpha]}{100},$$

where α is the rotation in degrees of a solution containing c g of substance in 100 ml of solution when measured in a column of length l decimetres and M = molecular weight. The rotation, α, is proportional to the difference $(n_l - n_r)$ in the refractive indices of the medium towards right- and left-handed circularly polarized light. Plane polarized light is the resultant of right- and left-handed circularly polarized components which, in an optically active medium, travel with unequal velocities and on emergence and re-combination produce plane-polarized light which has been rotated through an angle relative to the plane of the incident light.

In wavelength regions far removed from absorption bands, the rotatory dispersion varies monotonically with wavelength, increasing steadily as the wavelength of light decreases. When the medium under examination possesses one or more absorption bands which are optically active, anomalous rotatory dispersion is observed in the vicinity of these. The following sequence of changes in optical rotation is observed. On approaching the absorption band from the long wavelength side, the rotation begins to increase rapidly due to an increase in $(n_l - n_r)$ and reaches an extremum. The sign of the rotation depends on whether n_l is greater or less than n_r. For example, if $n_l > n_r$, left-handed circularly polarized light is relatively delayed in passing through the medium and this leads to a dextro-rotation of the plane-polarized resultant of the two circularly polarized components. Then the rotation drops to zero ($n_l = n_r$) and may reach another extremum in the opposite sense, where $n_r > n_l$. The absorption band has now been passed and the specific rotation drops to a more normal value and then, with further decrease in wavelength, begins to increase monotonically. These changes are illustrated in Fig. 22.5 for $(+)$ and $(-)[\text{Coen}_3]^{3+}$. This shows that, just as for the configurations of the enantiomers, their ORD curves are mirror images of each other.

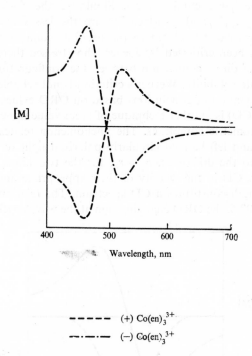

$----$ $(+)$ Co(en)$_3^{3+}$

$-\cdot-\cdot-$ $(-)$ Co(en)$_3^{3+}$

FIG. 22.5. Optical rotatory dispersion (ORD) curves for the enantiomers of $[\text{Coen}_3]^{3+}$.

An ORD curve does not make it possible to decide the configuration of a given enantiomer but, as the absolute configuration of $(+)[\text{Coen}_3]^{3+}$ is known, we may regard its ORD curve as diagnostic of this configuration. The significance of this is that any other optically active tris-chelate complex which shows the same type of ORD curve can be presumed to have the same configuration. For example, the ORD curves of the cations $(-)[\text{Iren}_3]^{3+}$ and $(-)[\text{Rhen}_3]^{3+}$ are similar[26] to that of $(+)[\text{Coen}_3]^{3+}$ and are therefore assigned the same configuration.

The $(+)[\text{Coen}_3]^{3+}$ ion has optically active absorption bands and its ORD curve is a combination of the individual curves arising from these absorptions. The band at 469 nm is associated with the electronic transition $^1A_{1g} \to {}^1T_{1g}$ typical of the d^6 ions Co(III), Rh(III) and Ir(III) in octahedral coordination and it is reasonable to expect that, for a given configuration, the ORD curves arising from this transition will be similar.

The variation of optical rotation with wavelength makes it clear that there is no significant relationship between absolute configuration and the sign of optical rotation at the sodium D line. For example, the negative rotation observed at this wavelength for $(-)[\text{Iren}_3]^{3+}$ is fortuitous and arises from the swamping of a small positive rotation by a large negative contribution from an optically active absorption band at shorter wavelengths.

The assignments made by comparisons of ORD curves are generally the same as those derived by applying Werner's original method for the correlation of configurations.[27] He proposed that the less soluble diastereoisomers of similar complexes with the same resolving agent have the same configuration. For example, $(+)[\text{Coen}_3]\text{Cl}$ $(+)$tartrate $5\text{H}_2\text{O}$ and $(-)[\text{Rhen}_3]\text{Cl}$ $(+)$tartrate $4\text{H}_2\text{O}$ are the less soluble diastereoisomers and are presumed to have the same configuration. Similarly, the $(+)$nitrocamphor salts of $(+)[\text{Cren}_3]^{3+}$, $(-)[\text{Rhen}_3]^{3+}$ and $(-)[\text{Iren}_3]^{3+}$ are the less soluble and so the configurations of all three ions can be related to the absolute configuration of $(+)[\text{Coen}_3]^{3+}$. Werner's proposal has been criticized by Jaeger who stressed there is no plausible justification for supposing a direct connection between the configurations of compounds and their relative or absolute solubility. Werner's criterion is not reliable on its own but it can provide a useful confirmation of conclusions based on ORD evidence.

Circular dichroism (CD) is another technique of great value in determining the absolute configuration of dissymmetric complexes. The phenomenon relates to the absorption by a substance of right- and left-handed circularly polarized light to different extents. It is directly proportional to the difference, $\varepsilon_l - \varepsilon_r$, for the two kinds of light. All optically active substances show CD in the vicinity of their appropriate absorption bands and its variation with wavelength constitutes a CD spectrum. The relationship between CD and ORD is shown in Fig. 22.6, the ORD changing sign at the wavelength of the CD extremum.

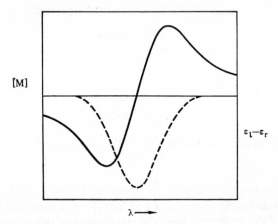

FIG. 22.6. The relationship between ORD (—) and CD (---).

The inversion of ORD and the CD in the vicinity of an absorption band were first discovered by Cotton[28] and each of the phenomena or their combination constitutes the Cotton effect. Figure 22.6 illustrates negative CD which is associated with a change in ORD from negative to positive on going from longer to shorter wavelengths whereas a positive CD band produces the opposite change.

The general principle which applies equally to ORD and CD is that 'two related optically active molecules have the same absolute configuration if they give a Cotton effect of the same sign in the absorption wavelength region of an electronic transition common to both molecules'.[29] The successful use of this principle hinges on the identification of at least one of the electronic transitions associated with the observed Cotton effect.

The CD and absorption spectra of $(+)[Coen_3]^{3+}$ are illustrated in Fig. 22.7 and are now considered in the light of the electronic transitions possible. For a low-spin Co(III) complex like this, four absorption bands are typically observed in the near infrared, the visible and the near ultraviolet regions. The metal ion is d^6 and, in octahedral coordination,

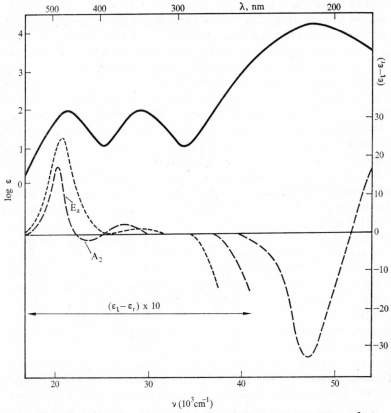

FIG. 22.7. The absorption spectrum (——) and CD (----) of $(+)[Coen_3]^{3+}$ in aqueous solution and the CD (........) of a single crystal of $2[(+)(Coen_3)Cl_3]$ $NaCl \cdot 6H_2O$ for radiation propagated along the optic axis (reproduced by permission of Professor S. F. Mason and amended from the original in *Mol. Phys.* **6**, 359 (1963)).

the ground state is $^1A_{1g}$. The absorption bands are due to spin-allowed and spin-forbidden transitions to higher energy levels. According to the Tanabe–Sugano diagram[30] for Co(III), there are two spin-allowed transitions:

(A) $^1A_{1g} \rightarrow {}^1T_{1g}$ and

(B) $^1A_{1g} \rightarrow {}^1T_{2g}$

and these are identified respectively with bands observed at 476 and 342 nm in $Co(NH_3)_6{}^{3+}$. Two spin-forbidden transitions

(C) $^1A_{1g} \rightarrow {}^3T_{1g}$ and

(D) $^1A_{1g} \rightarrow {}^3T_{2g}$

occur in the near infrared region and have very low intensities. All four excited states are triply degenerate and are split when the symmetry of the complex is lowered. When 'en' replaces NH_3 as ligand, the symmetry of the complex $[Coen_3]^{3+}$ becomes D_3 (compared with O_h for $[Co(NH_3)_6]^{3+}$). In D_3, the T_{1g} level splits into $^1A_2$ and ^1E and the $^1T_{2g}$ level into $^1A_1$ and ^1E. The $d \rightarrow d$ absorption bands should show evidence of this splitting but, in fact, the absorption spectrum of $[Coen_3]^{3+}$ in solution at room temperature closely resembles that of $[Co(NH_3)_6]^{3+}$, indicating that the band splitting in the former complex is too small to be observed.

In contrast, the CD spectrum clearly shows splitting in the region of band (A). Magnetic dipole selection rules predict the occurrence of two CD components, $^1A_1 \rightarrow {}^1A_2$ and $^1A_1 \rightarrow {}^1E_a$, arising from the splitting of $^1T_{1g}$ in the trigonal field. The observed CD bands (Fig. 22.7) are, in fact, residual wing absorptions resulting from the extensive cancellation of the rotational strengths of these two transitions.[31] The positive CD component is associated with the transition $^1A_1 \rightarrow {}^1E_a$. This conclusion is based on CD measurements on a single crystal of the double salt, $2 [(+)Coen_3Cl_3]NaCl6H_2O$, which, provided these are made with radiation directed along the optic axis of the crystal, record only the dichroism due to the transition involving E_a. This is because the C_3 axis of each complex ion is parallel to the optic axis of the crystal and whereas the component of A_2 symmetry has an electric and a magnetic moment directed along this axis, E_a has moments in a perpendicular direction to this axis. Figure 22.7 shows the single CD band observed in the solid state for this transition which allows the above identification. It is now possible to use this as the basis for other assignments using the general principle that other trigonal 'd^6 complexes have the same configuration as $(+)[Coen_3]^{3+}$ if the spin-allowed transition of lowest energy has an E component of positive rotatory power'.[31]

It has been pointed out[29] that the comparison of the general form of Cotton effect curves is not necessarily always a reliable guide to the assignment of configuration in cases where the major optical rotatory power is due to two or more electronic transitions which are degenerate in the corresponding symmetric complex. This is particularly so when the frequency differences between the components into which the degenerate transition is split in the dissymmetric molecule are small and variable in sign.

Configurations assigned to trigonal complexes on the above principle are generally in accord with those related by Werner's solubility method or by ORD comparisons. For example, the $S(C_3)$ configuration is common to those isomers of d^3 and d^6 metal tris-chelate complexes which form the less soluble halide $(+)$tartrate salts and those of the corresponding tris (malonato) or (oxalato) complexes forming the less soluble $(-)$strychnine salts.

Ion-Pair Formation

Werner observed that the molecular rotation of an optically active complex in solution is affected by the counter ions present. This has been confirmed in more recent ORD and CD studies and regarded as evidence for the formation of outer-sphere ion-pair complexes.

The measurements of the CD of $(+)[Coen_3]^{3+}$ in solutions of constant ionic strength and containing varying concentrations of an ion, L^-, which forms outer-sphere complexes, have been interpreted in terms of the CD spectra for the successively-formed complexes, $[Coen_3]L^{2+}$, $[Coen_3]L_2^+$, $[Coen_3]L_3$, etc. Larsson and Norman[32] have reported up to three complexes with salicylate ion and up to four with both thiosulphate and selenite. The changes in CD for the first two outer-sphere complexes involving SeO_3^{2-} are small but major changes occur with the formation of the third and fourth complexes. Stability constants for these outer-sphere complexes have been calculated from CD measurements and the CD of the complexes $[Coen_3]L_3$ (where L = salicylate, benzoate or diethylthiocarbamate) recorded after they have been extracted into organic solvents. When phosphate is the counter ion, large changes in CD are found to accompany the formation of a 1:1 complex. Another change in CD occurs in the ultraviolet region where a new band develops on ion-pair formation. This originates in a charge-transfer transition between the counter ion and the complexed metal ion.

The general pattern of change followed with regard to the long wavelength transition (A) is that most anions, whether multiply or singly charged, diminish and enhance respectively, the areas under the CD bands due to the E_a and A_2 components.

Although this evidence appears to be very convincing, Olsen and Bjerrum[33] have questioned whether complexes including more than one anion are really formed. On the basis of their studies of the ultraviolet absorption spectrum of $[Coen_3]^{3+}$ in aqueous solution containing thiosulphate or selenite ions, they concluded that, except possibly at very high anion concentration, the only two species present were the complex cation itself and its 1:1 complex with the counter ion. Olsen and Bjerrum suggest that the changes observed in the CD of such solutions could be associated with the equilibrium which exists in solution between the four conformers of each optical isomer of $[Coen_3]^{3+}$.

References

1. JÖRGENSEN, S., *J. prakt. Chem.* [2], **39**, 8 (1889).
2. GROSSMANN, H., and SCHÜCK, B., *Ber.* **39**, 1899 (1906).
3. *Inorganic Syntheses*, vol. VI, p. 183.
4. WERNER, A., *Ber.* **45**, 121 (1912).
5. *Inorganic Syntheses*, vol. VI, p. 186.
6. DWYER, F. P., and SARGESON, A. M., *J. Phys. Chem.* **60**, 1331 (1956).
7. VAUGHN, J. W., MAGNUSON, V. E., and SEILER, G., *Inorg. Chem.* **8**, 1201 (1969).
8. COREY, E. J., and BAILAR, J. C., Jr., *J. Amer. Chem. Soc.* **81**, 2620 (1959).
9. ROSENBLATT, F., and SCHLEEDE, A., *Ann.* **505**, 54 (1933).
10. KOBAYASHI, M., *J. Chem. Soc. Jap.* **64**, 648 (1939).
11. NAKATSU, K., SAITO, Y., and KUROYA, H., *Bull. Chem. Soc. Jap.* **29**, 428 (1956).
12. SAITO, Y., NAKATSU, K., SHIRO, M., and KUROYA, H., *Bull. Chem. Soc. Jap.* **30**, 795 (1957).
13. DWYER, F. P., MACDERMOTT, T. E., and SARGESON, A. M., *J. Amer. Chem. Soc.* **85**, 2913 (1963).
14. MACDERMOTT, T. E., *Chem. Comm.* 223 (1968).
15. GOLLOGLY, J. R., and HAWKINS, C. J., *Inorg. Chem.* **9**, 576 (1970).
16. NISHIKAWA, S., and MATSUKAWA, K., *Proc. Imp. Acad. (Tokyo)* **4**, 96 (1928).
17. COSTER, D., KNOL, K. S., and PRINS, J. A., *Z. Physik.* **63**, 345 (1930).
18. PEERDEMAN, A. F., VAN BOMMEL, A. J., and BIJVOET, J. M., *Proc. Acad. Sci. Amst.* B **54**, 16 (1951).
19. SAITO, Y., NAKATSU, K., SHIRO, M., and KUROYA, H., *Acta Cryst.* **8**, 729 (1955).
20. SAITO, Y., IWASAKI, H., and OTA, H., *Bull. Chem. Soc. Jap.* **36**, 1543 (1963).
21. FAY, R. C., and PIPER, T. S., *J. Amer. Chem. Soc.* **84**, 2303 (1962).
22. SPEES, S. T., Jr., DURHAM, J., and SARGESON, A. M., *Inorg. Chem.* **5**, 2103 (1966).
23. YONEDA, H., and MORIMOTO, Y., *Bull. Chem. Soc. Jap.* **39**, 2180 (1966).
24. YONEDA, H., EMERSON, M. T., and MORIMOTO, Y., *Inorg. Chem.* **8**, 2214 (1970).
25. YANO, S., ITO, H., KOIKA, Y., FUJITA, T., and SAITO, K., *Bull. Chem. Soc. Jap.* **42**, 3184 (1969).
26. MATHIEU, J. P., *J. Chim. Phys.* **33**, 78 (1936).
27. WERNER, A., *Bull. Soc. Chim. France* **11**, 1 (1894).
28. COTTON, A., *Ann. Chim. Phys.* [7], **8**, 347 (1896).
29. MCCAFFERY, A. J., MASON, S. F., and BALLARD, R. E., *J. Chem. Soc.* 2883 (1965).
30. TANABE, Y., and SUGANO, S., *J. Phys. Soc. Jap.* **9**, 753, 766 (1954).
31. MCCAFFERY, A. J., and MASON, S. F., *Mol. Phys.* **6**, 359 (1963).
32. LARSSON, R. L., and NORMAN, B. J., *J. Inorg. Nucl. Chem.* **28**, 1291 (1966).
33. OLSEN, I., and BJERRUM, J., *Acta Chem. Scand.* **21**, 1112 (1967).

CHAPTER 23

COPPER(II) NITRATE

Introduction

Our knowledge of the physical and chemical properties of inorganic compounds is determined to a large extent by the use of water as the solvent in their preparation and of aqueous media for studying their behaviour in solution. There are two chief reasons for this: firstly, the availability of water above all other solvents; secondly, its high dielectric constant which means it is eminently suitable as a solvent in which to perform ionic reactions. The nature of water as a coordinating, polar molecule does mean, however, that it does not behave as an inert solvent but influences, often profoundly, the progress and products of a reaction. Thus the solvation of cations by water molecules means that, in most cases, the products isolated from aqueous solution are the hydrated compounds. For example, the familiar blue colour of copper(II) salts in aqueous solutions is due to the presence of hydrated copper(II) ions. These are also present in blue copper sulphate pentahydrate prepared by crystallization from aqueous solution. This salt can, however, be dehydrated by heating to give the white anhydrous salt. Many other hydrates can be thermally dehydrated but in some cases heating causes decomposition as well. Then it becomes necessary to make the anhydrous salt by alternative means, for example, the chemical dehydration of hydrated metal chlorides by thionyl chloride. More generally, anhydrous compounds are directly synthesized by performing the appropriate reactions in non-aqueous solvents.

Copper(II) nitrate is one of a number of anhydrous metal nitrates which have been intensively studied in recent years. The structure of this compound is now well established and a knowledge of it provides fresh insight into the bonding which occurs between a metal and the nitrato-group, $-NO_3$.

Preparation

Copper(II) nitrate has been prepared by a variety of methods.[1] For example, reaction between freshly-precipitated copper(II) oxide and dinitrogen tetroxide in a sealed Pyrex tube at 87°C leads to the formation of a green solid,[2] the analysis and properties of which show it to be the addition compound, $Cu(NO_3)_2 \cdot N_2O_4$. On heating this between 90° and 140° at very low pressure, copper(II) nitrate itself is obtained. The addition compound is also produced in the direct reaction of metallic copper with a mixture of dinitrogen tetroxide and ethyl acetate. The presence of ethyl acetate serves to raise the dielectric constant of the solution and therefore to promote reaction.

Metal nitrates have also been synthesized[3] by the reaction, at ordinary temperatures, between dinitrogen pentoxide, N_2O_5, and the metal, its anhydrous chloride or its hydrated nitrate. Dinitrogen pentoxide can be used on its own and this is advantageous in certain cases. The addition of a polar solvent to N_2O_4 to promote reaction can complicate matters because the second solvent itself is likely to coordinate strongly to the metal and make it more difficult to isolate the compound required in an unsolvated form.

Anhydrous copper(II) nitrate is a deep-blue crystalline solid which melts at 255–256°C. It is very deliquescent so it must always be manipulated in a dry-box.

The solid sublimes to a vapour containing monomeric copper nitrate molecules. Vapour pressure measurements[1] have shown that the vapour is stable up to 226°C, at which temperature decomposition to oxygen and nitrogen dioxide begins. Molecular weight determinations have shown that the nitrate is present as the monomer in the vapour at 217°C. This has been confirmed by mass spectrometry.[4] No copper-containing ions with masses greater than that of $Cu(NO_3)_2{}^+$ have been detected.

A careful study of the experimental conditions for making copper(II) nitrate has established that two distinct solid forms, known as α and β, exist.[5] The decomposition of $Cu(NO_3)_2 \cdot N_2O_4$ *in vacuo* at about 100°C, gives the α-form as a powder. Slow sublimation of this in dry air converts it to single crystals containing some β-form, the proportion of this increasing with the rate of sublimation. Sublimation *in vacuo* at 200°C produces β-form only. The $\alpha \rightarrow \beta$ transition is rapid above 150°C.

The two forms can be distinguished by their infrared spectra.[6] As shown in Table I, the spectrum of the β-form is the more complex of the two and is characterized particularly by a strong band at 1580 cm^{-1}.

The existence of more than one form of copper nitrate is confirmed by the different magnetic moments reported for samples prepared under different conditions. The deep blue-green nitrate prepared according to the method of Addison and Hathaway[1] has a magnetic moment of 1·78 B.M. at ordinary temperature.[7] This value is slightly above the spin-only value of 1·73 B.M. for a d^9 ion. The moment decreases monotonically to 1·59 B.M. at −190 °C. The magnetic properties are typical of solids in which there are weak antiferromagnetic interactions resulting, in this case, from superexchange operating through the bridging nitrato groups.

A sample of copper nitrate prepared from the deep blue-green compound by vacuum pumping and resublimation[8] gave $\mu_{eff} = 1\cdot94$ B.M. The difference between this and the above value has been attributed to some structural rearrangement caused by the extra treatment given to this sample. Such rearrangement could destroy the superexchange occurring through the nitrato-groups, leading to a higher value for the magnetic moment.

Structure

A thorough investigation of the structure of copper(II) nitrate in the vapour state by electron diffraction has been made by LaVilla and Bauer.[9] From the diffraction patterns obtained, the positions of maxima and minima were compared with theoretical intensity curves calculated for the three symmetric models regarded as feasible (Fig. 23.1). The best correlation between experimental and theoretical curves was found for the model in which the nitrato group is bidentate (Fig. 23.1b). The length of the bond between nitrogen and the coordinated oxygen is 0·130 nm, the Cu,O bond length is 0·200 nm, and the Cu,N interatomic distance is 0·230 nm. The bond angles calculated for Fig. 23.1b are O–N–O = 120 ± 2° and O–Cu–O = 70°.

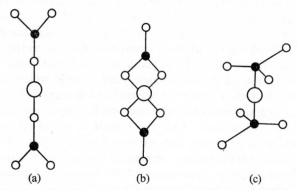

(a)   (b)   (c)

Fig. 23.1. Possible structures for the copper(II) nitrate molecule.

The orientation of the nitrato groups relative to each other cannot be established by electron diffraction for there is equally good agreement with experimental data whether the atoms of each nitrato group are regarded as co-planar with the rest of the molecule or individually co-planar with the copper atom but with the plane of one NO_3 group rotated about the N–Cu–N axis to give a tetrahedral configuration of the nearest oxygens around the metal. However, the Cu,O bond lengths in other compounds where the oxygen atoms are in a square-planar arrangement around the metal are within the range 0·190 to 0·200 nm and this supports the belief that the metal atom in the copper nitrate molecule is also in square planar coordination.

Confirmation of the molecular structure has come from a study of the E.S.R. spectrum of copper(II) nitrate, trapped at −269°C in a neon matrix.[10] The matrix was prepared on a sapphire rod, ground flat along most of its length so that a sample supported on it could be turned through known angles relative to the magnetic field in the microwave cavity of the E.S.R. spectrometer. The apparatus was specially designed to trap molecules which could only be produced by the vaporization of a solid phase at a high temperature.

The derivative E.S.R. spectrum obtained for two orientations of the sample in the magnetic field is shown in Fig. 23.2. In one case, the flat face of the sapphire rod is perpendicular to the field and in the other, it is parallel to the field. The spectrum can be interpreted by considering the central copper atom to be chiefly under the influence of the ligand field due to the four adjacent oxygen atoms. The symmetry of the group of atoms is D_{4h} and, in this case, g, the spectroscopic splitting factor, is anisotropic and characteristically has two different values, $g_{||}$ and $g_{\perp}$, which represent g when the z-axis (perpendicular to and passing through the centre of the plane containing the five atoms) of the molecule is respectively parallel to and perpendicular to the external magnetic field. Coupling of the electron spin with the spin of the ^{63}Cu nucleus, for which $I = 3/2$, produces hyperfine splitting into four $(2I + 1)$ lines for each g value.

In Fig. 23.2a, the four positively-directed lines at low fields are attributed to $g_{||}$. The doublet pattern at the lowest field arises from the presence of two different copper nuclei, ^{63}Cu and ^{65}Cu, which have equal spins but slightly different magnetic moments. These low-field lines almost completely disappear when the flat sapphire faces are parallel to

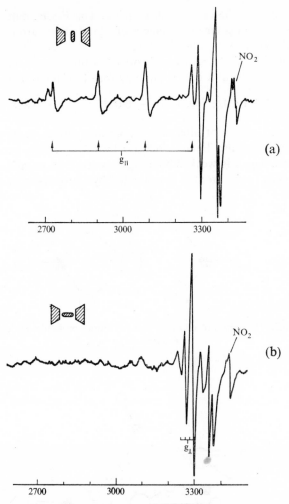

FIG. 23.2. The E.S.R. spectrum of copper(II) nitrate in a neon matrix at $-269°C$ observed (a) with the flat face of the rod perpendicular to the magnetic field, (b) with the flat face of the rod parallel to the magnetic field (reproduced with permission from Kasai, P. H., Whipple, E. B., and Weltner, W., *J. Chem. Phys.* **44**, 2581 (1966)).

the magnetic field (Fig. 23.2b). This indicates that the copper nitrate molecules are preferentially oriented so that their planes are parallel to the flat sapphire faces.

The four stronger lines, positively and negatively directed, at higher fields are due to $g_\perp$. These show some decrease in the case where the sapphire faces are perpendicular to the magnetic field. Background signals are observed due to the presence of some dinitrogen tetroxide molecules in the matrix.

Calculated g values are:

$$g_\parallel = 2\cdot2489 \pm 0\cdot0003; \quad g_\perp = 2\cdot0522 \pm 0\cdot0005$$

These values are very close to those for bis(acetylacetonato)copper(II),[11] $g_\parallel = 2\cdot266$ and $g_\perp = 2\cdot053$. In this complex, the acetylacetonato groups are bonded to the metal through oxygen atoms located at the corners of a square. The E.S.R. data on copper nitrate are consistent with a square planar arrangement of ligand atoms around the metal and so the structure of the molecule is as shown in Fig. 23.3.

The X-ray analysis[5] of the α-form of copper nitrate shows this belongs to the space group $Pmn2_1$ and that the unit cell, containing four $Cu(NO_3)_2$ groups, has dimensions:

$$a = 1 \cdot 112 \pm 0 \cdot 002; \ b = 0 \cdot 505 \pm 0 \cdot 001; \text{ and } c = 0 \cdot 828 \pm 0 \cdot 002 \text{ nm}$$

The solid contains chains of alternate copper and nitrato groups bonded strongly together. These chains are also linked sideways by strong bonds between alternate copper and nitrato groups. There is weaker bonding between each copper atom and two further oxygens so that each metal atom is coordinated by six oxygens in a distorted octahedral

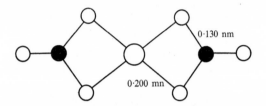

FIG. 23.3. The structure of the copper(II) nitrate molecule as determined by electron diffraction.

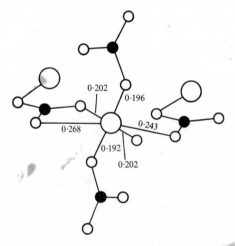

FIG. 23.4. The environment of copper atoms in α-copper(II) nitrate.

arrangement. This is shown in Fig. 23.4. The oxygen atom at a distance of $0 \cdot 268$ nm and one of the strongly-bound oxygens at $0 \cdot 202$ nm are in the same NO_3 group. This therefore behaves as an unsymmetrical bidentate ligand towards copper. The other nitrato groups act as bridges for pairs of copper atoms. No structural data on β-$Cu(NO_3)_2$ appear to have been published yet so it is not possible to give a full description of the structures of the two solid forms of this compound.

Infrared Spectrum

To progress beyond a purely qualitative use of the infrared spectrum for identification purposes to the assignment of bands to particular vibrations presents certain difficulties. Thus, assuming that the metal atom and the nitrogen and oxygen are all co-planar, there are three different ways in which the nitrato group can behave with respect to the metal atom. These are illustrated in Fig. 23.5, which represents (a) monodentate bonding to the metal, (b) bidentate bonding and (c) bridging action of the nitrato group in bonding two metal atoms. All of these structural models belong to the same point group, C_{2v}, and they give rise to the same number of fundamental vibrations in the spectrum. This cannot therefore provide unequivocal evidence on the mode of attachment of the nitrato group to the metal.

The band assignments in the infrared spectra of various anhydrous metal nitrates have been made on the basis of a monodentate attachment of the nitrato group to the metal.

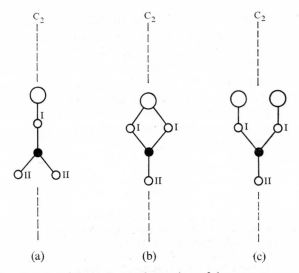

(a)   (b)   (c)

FIG. 23.5. Possible nitrato-metal groupings of the same symmetry.

TABLE I

Infrared Absorption Bands of α- and β-copper(II) Nitrate (frequency in cm^{-1})

α		β
1510		1580
1502	ν_1	1526
1478		1490
1435		1321
1300	ν_4	1254
1049	ν_2	1035
1015		1013
796	ν_3	1004
759	ν_5	790
746		777
712	ν_6	764
		735
		718

They are summarized in Table II. The complexity of the spectra of the α- and β-forms of $Cu(NO_3)_2$ indicates that there is probably more than one kind of nitrato group present in each case. Thus the strong band observed at 1580 cm^{-1} in the β-form probably originates from a bidentate or a bridging group.

TABLE II

Band Assignments in the Infrared Spectra of Metal Nitrates

Species	$\nu_1(A_1)$	$\nu_2(A_1)$	$\nu_3(A_1)$	$\nu_4(B_1)$	$\nu_5(B_1)$	$\nu_6(B_2)$
Assignment	NO_2^{II} sym. stretch	NO^I stretch	NO_2^{II} sym. bending	NO_2^{II} asym. stretch	NO_2^{II} asym. bending	Out of plane rocking
Frequency range (cm^{-1})	1290—1253	1034—970	739	1531—1481	713	800—781

Two novel features in the chemistry of copper(II) nitrate thus deserve special mention. Firstly, the existence of discrete molecules in the vapour state shows that the bonding within them confers appreciable thermal stability. Evidently, the nitrato group behaves as a strongly binding ligand towards copper under anhydrous conditions, in contrast to the relatively weak complexing which the nitrate ion shows in aqueous solution. Secondly, there are two distinct forms of the solid and this suggests that there are at least two different stereochemical arrangements of copper atoms and nitrato groups in a crystalline lattice. It is probable that these arise from the ability of the nitrato group to function in more than one way with respect to a metal atom.

References

1. ADDISON, C. C., and HATHAWAY, B. J., *J. Chem. Soc.* 3099 (1958).
2. FERRARO, J. R., and GIBSON, G., *J. Amer. Chem. Soc.* **75**, 5747 (1953).
3. FIELD, B. O., and HARDY, C. J., *J. Chem. Soc.* 4428 (1964).
4. PORTER, R. F., SCHOONMAKER, R. C., and ADDISON, C. C., *Proc. Chem. Soc.* 11 (1959).
5. WALLWORK, S. C., and ADDISON, W. E., *J. Chem. Soc.* 2925 (1965).
6. LOGAN, N., and SIMPSON, W. B., *Spectrochim. Acta* **21**, 857 (1965).
7. KOKOT, E., and MARTIN, R. L., *Chem. Comm.* 187 (1965).
8. BOUDREAUX, E. A., and MILLER, D. J., *Inorg. Nucl. Chem. Letters* **2**, 59 (1966).
9. LAVILLA, R. E., and BAUER, S. H., *J. Amer. Chem. Soc.* **85**, 3597 (1963).
10. KASAI, P. H., WHIPPLE, E. B., and WELTNER, W., *J. Chem. Phys.* **44**, 2581 (1966).

CHAPTER 24

ASTATINE

The periodic classification of the elements, which originated in the work of Mendeleev makes it possible to group these in logical sequence according to their chemical properties. Such a classification provides an essential framework for the study of inorganic chemistry and has, of course, a sound theoretical basis in terms of the electronic configurations of atoms. When the positions of all the naturally-occurring elements had been established, it became evident that below uranium (atomic number = 92) there are a number of 'missing' elements, none of which is found in nature. These include elements of atomic number 43 and 61 and some of atomic number higher than that (83) of bismuth. As one consequence of the research into the transuranic elements and nuclear fission carried out immediately before and during the Second World War, it has become possible to make isotopes of these 'missing' elements by the appropriate nuclear reactions. Their chemical properties have been studied, albeit only at tracer level for those nuclides which are intensely radioactive, and compared with those of stable elements in the same periodic group or series.

The halogens comprise a chemical family of outstanding importance and interest. One feature of their properties is that many of the trends observed in going down the group from fluorine to iodine are by no means regular and so it proved a particularly attractive problem to achieve the synthesis of the missing halogen, element 85, and to study its chemistry. All nuclides above ^{209}Bi are radioactive and so element 85 is expected to be unstable. Many isotopes of this have now been made and this expectation has been verified for all of them. In fact, the name astatine, given[1] to this element, is a derivative of the Greek word for unstable. (This nomenclature follows the pattern established for the other halogens, each being named by modifying the Greek word which denotes some characteristic property of the element in question.)

There is a number of formidable practical difficulties in studying the chemistry of an element like astatine. The main reason for this is that the longest-lived isotopes have half-lives of several hours only and are intensely radioactive. Investigations must therefore be carried out at tracer levels (where the molarity of solutions with respect to astatine is less than 10^{-10}) and often necessitates the use of carrier materials, generally iodine compounds. There appears to be no prospect of obtaining weighable amounts of astatine and, even if this were possible, their activity would be so great that the energy released would lead to intense local heating effects.

From its position in the Periodic Table following iodine, one might expect that the high oxidation states of astatine would be less stable than those of iodine and that the electropositive character of its lower oxidation states would be more marked than for iodine. A factor which complicates the direct comparison of the two elements is that the chemistry of elements at tracer levels often differs greatly in important respects from that at macro levels. For example, when dilute solutions (about 10^{-7} mol dm^{-3}) of iodide ion are oxidized under various conditions,[2] there are formed molecular iodine and three distinct compounds of iodine, two of which are extractable from aqueous sulphuric acid by organic solvents. At least some of these compounds are believed to be the products of reaction between traces of organic impurities and iodine in oxidized forms like HIO. Adsorption and radiocolloid formation are other phenomena important at low concentration which further complicate the study of astatine chemistry in solution.

Preparation

The first synthesis of astatine was reported[3] in 1940 and preliminary accounts[4,5] of its chemistry given. The bombardment of lead or bismuth targets with 5·13 pJ α-particles was observed to give rise to a number of products which were themselves α-emitters. One of these, with a half-life of about 7·5 hr, was identified as an isotope of astatine. Its chemical properties were described as being closer to those of polonium, its Group VI neighbour, than to those of iodine. For example, astatine can be co-precipitated with a sulphide and also precipitated from acid solutions by reducing agents like zinc and stannous chloride. The chemical nature of the astatine when precipitated thus cannot be described with certainty but these reactions suggest that the element is more metallic in character than iodine. It was noted also that astatine has some of the properties, such as volatility at comparatively low temperatures, which are characteristic of the heavier halogens. Thus when metallic bismuth is heated after α-irradiation, it loses most of its astatine activity before melting at 275°C.

A large number of astatine isotopes is now known. The preparation and decay characteristics of over 20 of these is described elsewhere.[6] The isotopes with mass numbers between 200 and 212 inclusive are all α-emitters and some are also able to decay by orbital electron capture. Those which have the longest half-lives and which are therefore most suitable for chemical studies are ^{211}At and ^{210}At (half-lives are 7·5 and 8·3 hr respectively). Isotopes of mass numbers between 213 and 219 are all short-lived α-emitters. Three of these occur as members of the uranium and actinium decay series.

Both ^{211}At and ^{210}At are products of the irradiation of bismuth with α-particles. When the energy of these is between 3·36 and 4·64 pJ, ^{211}At is formed in the nuclear reaction:

$$^{209}_{83}\text{Bi} + {}^{4}_{2}\text{He} \rightarrow {}^{211}_{85}\text{At} + 2{}^{1}_{0}\text{n}$$

When higher energy (>4·8 pJ) particles are used ^{210}At is also produced:

$$^{209}_{83}\text{Bi} + {}^{4}_{2}\text{He} \rightarrow {}^{210}_{85}\text{At} + 3{}^{1}_{0}\text{n}$$

The isotope ^{211}At decays by two routes: 41% by α-emission to ^{207}Bi ($t_{\frac{1}{2}} = 30$ y) and 59% by electron capture to ^{211}Po ($t_{\frac{1}{2}} = 1$ s) which decays by α-emission to ^{207}Pb. Therefore for each nucleus of ^{211}At which decays, one α-particle will be emitted, either by ^{211}At or ^{211}Po and so α-particle counting can be used to follow astatine through a chemical process.

Alternatively, ^{211}At can be determined by measurement of its X-radiation which accompanies the electron capture process leading to ^{211}Po.

The isotope ^{210}At decays almost entirely (99·9%) by electron capture to ^{210}Po, the remaining 0·1% by α-emission decaying to ^{206}Bi. The α-emission by ^{210}At is thus extremely small and the nuclide is best measured either by its X-radiation or γ-radiation (both are produced in the electron capture process). The polonium isotope, ^{210}Po ($t_{\frac{1}{2}} = 140$ d), decays by α-emission but, in view of its relatively long half-life, its contribution to the total activity of irradiated bismuth is quite small. This nuclide is also formed directly in the irradiation of bismuth:

$$^{209}_{83}\text{Bi} + {}^4_2\text{He} \rightarrow 2{}^1_0\text{n} + {}^1_1\text{H} + {}^{210}_{84}\text{Po}$$

and also, if the α-particle beam contains deuterium ions, by

$$^{209}_{83}\text{Bi} + {}^2_1\text{D} \rightarrow {}^1_0\text{n} + {}^{210}_{84}\text{Po}$$

A chemical separation of polonium from astatine can be effected during the processing of target material after irradiation, but if ^{211}At is present, its decay will necessarily lead to a build-up of polonium concentration during subsequent work and the α-activity due to this must always be taken into account.

The target material commonly used in astatine preparation is either bismuth metal or the oxide.[6,7] The metal itself can be fused on to another metal for support (gold and silver are among metals used for this purpose). One technique used for the irradiation of the oxide is to place this in holes drilled along one edge of a metal plate (aluminium or copper). The other edge of the metal plate is clamped between two cooled copper blocks and the whole assembly irradiated in the vacuum chamber of the cyclotron. The cooling is essential to reduce to a minimum the loss of astatine by evaporation as it is formed.

To separate astatine from irradiated bismuth or bismuth oxide, distillation and solvent extraction techniques are used.[8] The element itself can be readily removed by distillation, either *in vacuo*[9] or by sweeping the vapour in a stream of nitrogen gas through a series of cooled traps, and then collected on a cold finger. Collection can be directly on to the glass surface or on to a metal disc attached to the cold finger. The latter is preferable in view of the volatility of astatine from a glass surface. Even at room temperature, the loss of astatine is rapid with an estimated "half-life" of evaporation of approximately 1 hr.[9] In contrast, the rate of loss from gold or platinum surfaces is much less. The astatine is conveniently removed from the cold finger or metal disc by treatment with a suitable solvent.

Solvent extraction has proved to be particularly valuable in general studies of astatine as well as for the isolation of the element from irradiated material. For example, the bismuth target can be dissolved in nitric acid, the solution treated with concentrated HCl (to bring the molarity of this acid up to 9) and the astatine extracted with isopropyl ether. About 90% of the astatine activity is extracted into the ether phase. In the aqueous phase, the element is probably in the form of the chloro-complex $AtCl_2^-$ and is extracted therefrom as the solvated parent acid $HAtCl_2$. The behaviour parallels the extraction of iodine into ethers from hydrochloric acid solutions containing the complex ion ICl_2^-. Astatine can be back-extracted from organic solvents into alkaline aqueous solutions, probably because of conversion to hypoastatite, AtO^-. Alternatively, irradiated bismuth oxide is dissolved in aqueous perchloric acid which contains some iodine. The bismuth is then precipitated as phosphate and the astatine extracted from the aqueous filtrate by carbon tetrachloride or chloroform.[7] This gives an organic extract in which the astatine is presumed to be present as AtI. Back extraction into an aqueous phase is possible when this contains a reducing agent.

The activity of astatine can be measured by its α- or X-radiation. For α-counting, aqueous solutions are placed on silver or platinum discs and evaporated to dryness. Alternatively, astatine is co-precipitated with other materials. For example, tellurium carries down astatine when tellurous acid is reduced in hydrochloric acid solution. In the presence of iodine, astatine can be precipitated with an iodide from a reducing solution. Thus astatine can be co-precipitated with silver iodide after reduction of silver nitrate in nitric acid with sulphur dioxide. This appears to be due rather to the deposition of astatine on the surface of metallic silver rather than to the formation of silver astatide itself. Losses of astatine have been noted in this procedure and precipitation with PdI_2 has been recommended[10] as a more satisfactory procedure.

In addition to the bombardment of bismuth with α-particles, two other types of nuclear reaction have been used to prepare astatine. These are the bombardment of thorium or uranium with protons, deuterons or α-particles of very high energy (several hundred MeV) and the bombardment of lighter elements with heavy ions, as exemplified by the bombardment of gold targets with ions of ^{12}C, ^{13}C, ^{14}N and ^{16}O. Both kinds of reaction lead to the formation of the lighter isotopes of astatine, down to and including ^{200}At.

In view of the high instability of all known astatine isotopes, the element is not expected to occur in natural sources except possibly as short-lived isotopes present in certain minerals as the result of the radioactive decay of long-lived parent isotopes.

Radioactive nuclides are conveniently classified into four series: the thorium $(4n)$ series, which originates with ^{232}Th and ends with the stable nuclide, ^{208}Pb; the neptunium $(4n + 1)$ series, named after the longest-lived member, ^{237}Np, and ending with the stable nuclide, ^{209}Bi; the uranium–radium $(4n + 2)$ series, which begins with ^{238}U and ends with ^{206}Pb; and the actinium $(4n + 3)$ series, which starts with ^{235}U and ends with ^{207}Pb. In the thorium series, the term $4n$ signifies that the mass number is divisible by 4 whilst in the others, the mass number is divisible by 4 with remainders 1, 2 and 3 respectively. The $4n$, $(4n + 2)$ and $(4n + 3)$ series are all naturally occurring; the $(4n + 1)$ series has been artificially synthesized and only trace amounts of ^{237}Np and some of its daughter elements have been identified in natural sources. Thus about 2×10^{-12} parts by weight of ^{237}Np to one part of ^{238}U have been found in uranium ores.[11] The isotope ^{217}At $(t_{\frac{1}{2}} = 0\cdot02\ s)$ is a member of the $(4n + 1)$ series and its α-emission has been observed in the α-spectrum of ^{225}Ac samples isolated from these ores. The isotope ^{218}At $(t_{\frac{1}{2}} = 2\ s)$ is a branched-chain product in the uranium–radium series and ^{215}At $(t_{\frac{1}{2}} = 10^{-4}\ s)$ and ^{219}At $(t_{\frac{1}{2}} = 0\cdot9\ m)$ are branched-chain products in the actinium series. The last named is the longest-lived astatine isotope to occur in nature and the only one it is possible to isolate by chemical means.

Chemical Properties

At least five oxidation states of astatine have been identified: these are -1, 0, $+1$, $+3$ and $+5$. Evidence for these is derived from extraction and co-precipitation studies but some of the conclusions which have been drawn are open to question in view of the difficulties in interpreting results obtained with tracer concentrations of the element. Tentative reduction potentials given by Appelman and reported by Anders[12] are:

$$At^- \xrightarrow[At_2]{+0\cdot3} At_2 \xrightarrow{+0\cdot9} AtO \xrightarrow{+1\cdot5} AtO_3^-$$

Oxidation State -1

The astatide ion, At^-, is the best characterized state and closely resembles the iodide ion. It is usually formed in acid solution by the reduction of astatine(O) with SO_2 or zinc. No astatine is extractable from the reduced solution into carbon tetrachloride showing that At(O) is no longer present. The negative charge on the astatine-containing ion has been established by electromigration experiments.

Astatide is co-precipitated with silver iodide. This is an effective process for the initial removal of astatine from solution although the activity is not retained particularly well by the silver iodide precipitate. This suggests the astatine may not be chemically bound as silver astatide but is adsorbed on the surface of metallic silver.

The mobility of astatide relative to iodide ions has been measured. The ratio of the diffusion coefficients in 1% NaCl solution containing $1\cdot2 \times 10^{-3}\,\text{mol}\,\text{dm}^{-3}$ KI and $4 \times 10^{-3}\,\text{mol}\,\text{dm}^{-3}\,Na_2\cdot SO_3$ is $D_{I^-}/D_{At^-} = 1\cdot41$.

Oxidation State (0)

Astatine isolated from irradiated target material shows some properties which are expected for the zero oxidation state. These are the volatility of the element and the ease with which it may be extracted from nitric acid solution into organic solvents like benzene or carbon tetrachloride. Variable values of the distribution coefficient between the organic solvent and water have been observed, probably because of the reaction at tracer level of the element with impurities. If the organic phase containing At(O) is washed repeatedly with dilute nitric acid, the distribution coefficient approaches a limiting value (91 for carbon tetrachloride). The distribution coefficients for the halogens show a trend of increasing values from chlorine to iodine and so we would expect the value for astatine to be greater than that for iodine. For carbon tetrachloride, the maximum value of 91 for astatine may be compared with the corresponding coefficient for iodine of 85. The iodine figure is based on work at macro levels and a value of 31 has been reported[9] at tracer concentrations. This difference underlines the dangers of attempting comparisons between data obtained at two very different levels of concentration.

It is uncertain whether the extractable species in the above systems is At_2 or At. At the very low concentrations involved, it is rather unlikely that At_2 molecules are present. When a large excess of iodine is present, it is almost certain that the astatine is in the form of AtI molecules. Under these circumstances, the distribution coefficient between carbon tetrachloride and water is $5\cdot5$.[14]

When an alkaline solution ($0\cdot1\,\text{mol}\,\text{dm}^{-3}$ NaOH) of astatine is equilibrated with carbon tetrachloride or benzene, the distribution coefficient drops to a value near 1. The simplest explanation of this behaviour is that astatine disproportionates to astatide ions and an oxidized state. All the astatine activity in the alkaline solution is precipitated with silver iodide suggesting that, as well as At^- ions, the oxidized state of astatine can be removed from solution with this carrier.

Except for At(O) and interhalogen compounds, none of the other oxidation states extracts into organic solvents. This property forms the basis for experimental oxidation and reduction studies on astatine by measuring the changes in its distribution coefficient after treatment with different reagents.[9] For example, large decreases in distribution coefficient are observed after the addition of SO_2 or arsenious acid to an aqueous solution containing astatine. This is consistent with reduction to astatide. The reaction with oxidizing agents like persulphate or iodate again leads to smaller distribution coefficients. An oxidized form which co-precipitates with insoluble iodates is produced and is therefore presumed to be astatate, AtO_3^-. Astatine is also slowly oxidized by cold, concentrated nitric acid and by ferric salts. Reduction back to the zero oxidation state is effected readily by ferrous salts.

Oxidation State +1

Astatine is oxidized in nitric acid/dichromate solution to the univalent cation, At^+. Electromigration and ion-exchange studies[15,16] have confirmed that astatine carries a positive charge. At^+ co-precipitates, *inter alia*, with thallium (I) and silver dichromates, silver iodate and the tungsto- and molybdophosphates of caesium.

The At^+ ion forms chloro-complexes $AtCl$ and $AtCl_2^-$ in the presence of hydrochloric acid. The stability constants of these have been measured[17] by studying the distribution of astatine between a cation-exchange resin and an aqueous solution $0.5\,mol\,dm^{-3}$ in HNO_3, $0.005\,mol\,dm^{-3}$ in $H_2Cr_2O_7$ and $(4 \pm 2) \times 10^{-3}\,mol\,dm^{-3}$ in HCl. The following values were obtained:

$$\beta_1 = 7 \times 10^2\,mol^{-1}\,dm^3 \quad \text{and} \quad \beta_2 = 2.5 \times 10^5\,mol^{-2}\,dm^6.$$

The bis(pyridine) complexes of iodine in its +1 oxidation state are well known and analogous compounds have been made for bromine. Astatine should form similar complexes which, if astatine is more electropositive than iodine, should be more stable than the iodine complexes. The compound $I(py)_2ClO_4$ has been synthesized in the presence of astatine. The iodine complex is made by the addition of iodine to $Ag(py)_2ClO_4$ in chloroform solution. Silver iodide is precipitated and removed by centrifuging and filtration and $I(py)_2ClO_4$ (accompanied by astatine) is precipitated from the filtrate by adding ether. The ratio of astatine to iodine concentrations in the solid is about two or three times the average value in solution prior to complex formation. This indicates that astatine preferentially forms the pyridine complex and supports the hypothesis that astatine should occur in a positive oxidation state rather more readily than iodine.

Oxidation State +3

This appears to exist in aqueous solutions produced by the action of moderately strong oxidizing agents like bromine, ferric iron or vanadium(V) on astatine. An intermediate state of astatine is also produced by the reduction of astatine(V) with chloride. Evidence for the existence of +3 astatine is not conclusive and it is quite possible that these reactions lead to the formation of AtO^- (containing astatine in a +1 oxidation state).

Oxidation State +5

Astatate ions, AtO_3^-, are formed by the action of strong oxidizing agents on astatine solutions. They are quantitatively carried by silver, lead and barium iodates and lanthanum hydroxide. No activity is carried by potassium periodate and this suggests that a perastatate (AtO_4^-) is not formed in these reactions. The +5 state appears to be relatively stable and there is no evidence that +7 astatine can be made. We should certainly expect the highest oxidation state of astatine to be difficult to prepare in view of the position of the element in the Periodic Table.

Astatine in Organic Compounds

Various astatine-labelled organic compounds have been synthesized. These generally resemble the analogous iodine compounds. They are potentially of great utility in biological studies because of this similarity. For example, like iodine, astatine is selectively concentrated in thyroid tissue.

Aten[18] has pointed out that, to obtain stable astatine-containing organic compounds, the logical approach is to incorporate the astatine atom in molecules where the nature of the other groups present ensures that it carries a partial positive charge. Following this line of thought, Samson and Aten have prepared[19] astatoacetic acid, $AtCH_2COOH$, by reaction between astatide in solution containing iodide as carrier and iodoacetic acid:

$$ICH_2COOH + At^- \rightarrow AtCH_2COOH + I^-$$

They demonstrated that astatoacetic acid was formed by adsorption of the reaction products on an anion-exchange chromatographic column and the behaviour on elution with aqueous sodium nitrate. The astatine activity closely followed the elution of iodoacetic acid and could be clearly differentiated from astatide activity.

The labelling of organic compounds with radionuclides can be carried out using a gas chromatographic technique in which the radioactive material is incorporated in the stationary phase of a column and the compound to be labelled flows through the column in the gaseous phase. In the application of this method to the synthesis of n-alkyl astatides,[20] ^{211}At and iodine in potassium iodide solution is adsorbed on Kieselguhr packed into a copper column. A sample of n-alkyl iodide or bromide is injected on to the top of this column and, as the volatile material passes down the column, exchange leads to the formation of the corresponding astatine compound. Gas chromatographic analysis of the products has shown in each case studied that the activity is associated with a compound less volatile than the alkyl halide used as a starting material. This confirms the synthesis of an n-alkyl astatide.

References

1. CORSON, D. R., MACKENZIE, K. R., and SEGRÈ, E., *Nature* **159**, 24 (1947).
2. KAHN, M., and WAHL, A. C., *J. Chem. Phys.* **21**, 1185 (1953).
3. CORSON, D. R., and MACKENZIE, K. R., *Phys. Rev.* **57**, 250 (1940).
4. CORSON, D. R., MACKENZIE, K. R., and SEGRÈ, E., *Phys. Rev.* **57**, 459 (1940).
5. CORSON, D. R., MACKENZIE, K. R., and SEGRÈ, E., *Phys. Rev.* **57**, 1087 (1940).
6. HYDE, E. K., PERLMAN, I., and SEABORG, G. T., *The Nuclear Properties of the Heavy Elements*, Prentice-Hall, New Jersey, 1964.
7. ATEN, A. H. W., DOORGEEST, T., HOLLSTEIN, U., and MOEKEN, H. P., *Analyst* **77**, 774 (1952).
8. NEUMANN, H. M., *J. Inorg. Nucl. Chem.* **4**, 349 (1957).
9. JOHNSON, G. L., LEININGER, R. F., and SEGRÈ, E., *J. Chem. Phys.* **17**, 1 (1949).
10. ATEN, A. H. W., VAN RAAPHORST, J. G., NOOTEBOOM, G., and BLASSE, G., *J. Inorg. Nucl. Chem.* **15**, 198 (1960).
11. PEPPARD, D. F., MASON, G. W., GRAY, P. R., and MECH, J. F., *J. Amer. Chem. Soc.* **74**, 6081 (1952).
12. ANDERS, E., *Ann. Rev. Nucl. Sci.* **9**, 203 (1959).
13. JOHNSTON, M., ASLING, C. W., DURBIN, P. W., and HAMILTON, J. G., University of California Radiation Laboratory, UCRL-3013, 35 (1955).
14. APPELMAN, E. H., Chemical Properties of Astatine, UCRL-9025, University of California (1960).
15. AN, FU-CHUN, NORSEEV, Y. V., CHAO, T-N, and KHALKIN, V. A., *Radiokhimiya* **5**, 351 (1963).
16. AN, FU-CHUN, KRYLOV, N. G., NORSEEV, Y. V., CHAO, T-N, and KHALKIN, V. A., *Akad. Nauk. SSSR* **80** (1965) (see *Chem. Abs.* **63**, 7695f).
17. NORSEEV, Y. V., and KHALKIN, V. A., Joint Inst. Nucl. Im. Rep. JINR-P-12, 3529 (1967) (see *Chem. Abs.* **69**, 70669y).
18. ATEN, A. H. W., *Adv. Inorg. Radiochem.* **6**, 207 (1966).
19. SAMSON, G., and ATEN, A. H. W., *Radiochim. Acta* **9**, 53 (1968).
20. SAMSON, G., and ATEN, A. H. W., *Radiochim. Acta* **12**, 55 (1969).

APPENDIX

INFRARED AND RAMAN SPECTRA

A molecule or polyatomic ion absorbs energy in the infrared region of the electromagnetic spectrum causing transitions from lower to higher vibrational energy levels. Normal or fundamental vibrational frequencies correspond with transitions from the ground to the first excited state. If the absorbing species is considered as a harmonic oscillator, the selection rule obeyed for absorption is $\Delta v = +1$, where Δv is the change in vibrational quantum number. As most molecules and ions are not perfect harmonic oscillators, this rule generally breaks down and transitions corresponding with $\Delta v = 2, 3$, etc., are also observed. When $\Delta v = 2$, the absorption frequency is approximately twice that of the fundamental frequency; when $\Delta v = 3$, it is approximately three times. These transitions are known respectively as the first and second overtones. Their absorption intensity is much less that that of the corresponding fundamental band. When two or more fundamental vibrations are simultaneously excited, then combination bands, for which the frequency is approximately the sum of the component frequencies, are observed.

The number of normal or fundamental vibrations possible in a polyatomic species can simply be expressed in the following way. The motion of each atom has three degrees of freedom, corresponding with movement in the x, y and z directions. For a molecule containing N atoms, there are $3N$ degrees of freedom. As three of these are associated with translatory movement and, in the case of non-linear molecules, three with rotation, the number of fundamental vibrations theoretically possible is $(3N - 6)$. For a linear poly-atomic molecule there are only finite moments of inertia about the two axes perpendicular to the molecular axis and hence only two degrees of rotational freedom. The number of vibrations possible in this case is $(3N - 5)$. Some of the fundamental vibrations are degenerate in many molecules and then the number of frequencies actually observed is smaller than indicated by these formulae.

Vibrations within a molecule are of different types, depending on the type of movement of the atoms. Thus there are 'stretching' vibrations, which cause bond lengths to alter, and there are 'deformation' vibrations which cause bond angles to change. Each vibration is not completely independent of movements occurring in the rest of the molecule and will be affected to some extent by these. There is always some mixing of one vibration with others but it is often convenient to identify each vibration as predominately of one type or the other.

Not all normal vibrations are active in the vibrational spectrum. Those allowed are expressed by two important selection rules. (1) A vibrating molecule absorbs infrared radiation only if there is a change in dipole moment as it vibrates. (2) A vibration is active in the Raman spectrum only when there is a change in the polarizability of the molecule associated with the vibration. The intensity of a band in the infrared spectrum depends on the magnitude of the dipole moment change. The intensity of a Raman band is related to the polarizability of the vibrating atoms.

Normal vibrational modes are defined by their symmetry properties. In group theory, the symmetry or point group to which a molecule belongs is described by its irreducible representations and each vibrational mode has a symmetry corresponding to one of these representations.

Irreducible representations and vibrations are distinguished, according to Mulliken's notation, in the following manner. Those which are symmetric with respect to a particular n-fold symmetry axis, C_n (usually the axis of highest n), are designated A. If antisymmetric with respect to this axis, they are labelled B. The symbols E and T are used to designate respectively two- and three-fold degenerate irreducible representations (vibrations). Subscripts 1 and 2 refer respectively to symmetric and antisymmetric irreducible representations relative to a C_2 axis perpendicular to the principal axis. The superscripts ′ and ″ refer to irreducible representations which are respectively symmetric or antisymmetric with reference to a horizontal plane of symmetry. Where there is a centre of inversion, the subscript g designates irreducible representations which are symmetric with respect to inversion and the subscript u is used for those which are antisymmetric.

The complete vibrational analysis of a molecule must be based normally on its infrared and its Raman spectra because of the inactivity of some of the fundamental vibrations in one or other of these. An important special case is that of a molecule which has a centre of symmetry and then the *rule of mutual exclusion* is obeyed. In other words, the infrared and Raman spectra have no absorption bands in common. This can be a valuable aid in the determination of molecular symmetry for, if an absorption band of the same frequency appears in both the Raman and infrared spectra, it follows that the molecule has no centre of symmetry.

Further valuable information is obtained from studying the degree of polarization of Raman lines because this makes it possible to associate these lines with particular kinds of vibration. We may consider the incident light reaching the molecule along a certain direction, say the x-axis. The scattered Raman radiation is observed at right angles to this, for example, along the y-axis. This scattered radiation is polarized and the intensities of its components, $I_{\parallel}$ and $I_{\perp}$ (respectively parallel and perpendicular to the x-axis) are different. The ratio if these, $I_{\perp}/I_{\parallel}$, defines the depolarization ratio, φ. The Raman lines for which $\varphi = 6/7$ are referred to as depolarized and correspond with molecular vibrations which are unsymmetric. Raman lines for which $0 < \varphi < 6/7$ are said to be polarized and correspond with totally symmetric molecular vibrations.

For details of the principles of infrared and Raman spectroscopy, the reader should consult appropriate texts.[1,2]

References

1. WALKER, S., and STRAW, H., *Spectroscopy*, vol. 2. Science Paperbacks (1967).
2. HERZBERG, G., *Infrared and Raman Spectra of Polyatomic Molecules*. D. Van Nostrand, Princeton, New York, 1945.

INDEX